Radionuclide Techniques in Clinical Investigation

Medical Physics Handbooks
Other books in the series

Series Editor: **Professor J M A Lenihan**
Department of Clinical Physics and Bio-Engineering
West of Scotland Health Boards, Glasgow

Medical Physics Handbooks 12

Radionuclide Techniques in Clinical Investigation

P W Horton
Armed Forces Hospital,
Riyadh, Saudi Arabia

Adam Hilger Ltd, Bristol
in collaboration with the
Hospital Physicists' Association

British Library Cataloguing in Publication Data

Horton, P W
    Radionuclide techniques in clinical investigation.
    (Medical physics handbooks; 12 ISSN 0143-0203).
    1. Radiology, Medical
    I. Title II. Series
    616.07'57   R895

    ISBN 0-85274-503-6

Published by Adam Hilger Ltd,
Techno House, Redcliffe Way, Bristol BS1 6NX

The Adam Hilger book-publishing imprint is owned by
The Institute of Physics

Printed in Great Britain by
Page Bros (Norwich) Ltd

Contents

Contents

Preface

Since their introduction more than thirty years ago, radioactive compounds have developed into an important tool in clinical diagnosis and research. Initially introduced to delineate metabolic pathways, both physiological and pathological, they are now also widely used as organ imaging agents. Both uses are practiced in the modern speciality of *nuclear medicine*. This monograph deals mainly with the first of these two applications. Its purpose is to describe the principles underlying the radionuclide tracer methods used to investigate clinical conditions. These principles are illustrated by reference to their common applications. However, such applications should not be regarded as comprehensive as new applications are always being devised in this innovative area of medical research. The production of safe and efficacious radioactive compounds and the practical measurement of radioactivity *in vivo* and *in vitro* are also described.

Two main characteristics have contributed to the very successful utilisation of radioactive compounds in clinical studies. First, modern techniques of measuring radioactivity are convenient. Where γ-ray emitting radionuclides are appropriate, their ease of detection external to the body allows painless and non-invasive means of investigation. Where biological samples, e.g. plasma or urine, are taken for assay, measurement of their radioactivity *in vitro* is direct and avoids the technical complexities of chemical analysis. Second, since radioactively labelled compounds possess the same chemical properties as the corresponding stable compound, they demonstrate the same behaviour *in vivo*. Therefore, they follow the same path through the body as the stable compound; this feature led to the early use of the term 'radioactive tracers' as a collective name. Radioactivity measurements typically have a sensitivity 10^4 times that of chemical analysis. This permits the administration of very small quantities of material for diagnostic tests and ensures that the metabolic process under investigation is not perturbed by the diagnostic procedure.

The tracer is distinguished by its radioactivity from the corresponding chemical material already in the system. Chemical assays necessarily measure the total amount of material present in a system or its parts and these may not change in spite of some kinetic process taking place if that process is in dynamic equilibrium. Measurement of the radio-activity therefore allows the movements in a system to be followed and this principle has been exploited in the following ways in physiological and pathological research and in diagnostic tests which have followed the research.

(a) The measurement of compartment or pool size by the dilution of the tracer in a compartment of the body. This relies upon equilibrium measurements.

(b) The measurement of absorption into the body, usually from the gastro-intestinal tract, and the loss of material from the body.

(c) The localisation of tracer in the body in particular organs. This technique is most important for organ imaging and a number of physio-logical mechanisms are exploited. These latter include concentration of a metabolite, phagocytosis, diffusion, cell sequestration, and capillary blockade.

(d) The measurement of rate factors for the passage of tracer from one body pool or compartment to another and the sojourn times of the tracer in these compartments.

A combination of measurements (b), (c) and (d) provides a complete quantitative description of the handling of the tracer in the body.

P W Horton
Riyadh, 1981

Acknowledgments

The composition of this handbook inevitably reflects the author's experience in nuclear medicine over the past twelve years. This experience was gained in the course of collaboration with scientific, clinical and technical colleagues in the Western Infirmary, Glasgow, the Department of Clinical Physics and Bio-engineering of the West of Scotland Health Boards, the Radionuclide Topic Group of the Hospital Physicists' Association and with many others. They are too numerous to mention by name but I would like to record my gratitude to them all for their help in scientific and technical matters and for their friendship. I am also indebted to Mrs Carol Downes, Miss Anne Wotherspoon and Miss Anne Stronach who helped prepare the typescript in very different parts of the world, and to Miss Margaret Finlayson who helped with the diagrams.

Glossary

Becquerel. The SI unit of radioactivity. One becquerel (Bq) is one nuclear transformation per second.

Carrier-free. A term used to describe a radiochemical containing only the radionuclide and none of the stable isotopes of the same element.

Curie. The old unit of radioactivity. One curie (Ci) is 3.7×10^{10} transformations per second (i.e. 3.7×10^{10} Bq).

Enriched material. Material containing an element in which the abundance of one of the isotopes has been increased over its natural abundance.

Half-life (biological). The time in which the radioactivity in a defined part of the body falls to half its initial value due only to biological redistribution, (i.e. after allowance for radioactive decay).

Half-life (physical). The time in which the quantity of a radionuclide decays to half its initial value.

Labelled compound. A chemical compound in which one or more of the atoms in a proportion of the molecules is replaced by a radionuclide. An intrinsically labelled compound is one in which one of the atoms of the molecule is replaced by a radionuclide of the same element.

Lyse. To destroy cells or tissues by pathological or chemical processes.

Radioactivity. The property of certain nuclides of emitting radiation during spontaneous transformation of their nuclei. Often colloquially termed 'activity'.

Radiochemical. As for labelled compound (above).

Radiochemical purity. The proportion of the total radioactivity in a radiochemical that is present in the stated chemical form.

Radioisotope. A radioactive species having the same atomic number but different mass numbers.

Radionuclide. A radioactive species of atom characterised by its mass number, atomic number *and* nuclear energy state.

Radionuclidic purity. The proportion of the total radioactivity in a radiochemical which is in the form of the stated radionuclide.

Radiopharmaceutical. A radiochemical in a suitable formulation for safe administration to persons, either orally or parenterally.

Senescence. The clearing of old red blood cells by the spleen.

Spallation reaction. A nuclear reaction in which several particles result from a collision.

Specific activity. The radioactivity per unit mass of an element or compound containing a radionuclide.

1 Production of Radionuclides

1.1 Introduction

Inspection of a chart of nuclides shows that there are about 1570 radionuclides with half-lives ranging from 10^{-6} to 10^{14} seconds, the most common half-life being about one hour. About 40 radionuclides are in common clinical internal use at the present time; their half-lives range from 13 seconds to 270 days. About two thirds of these radionuclides are produced in nuclear reactors, about one third in medical cyclotrons. A small number are produced by a 'generator' relying on the disintegration of a radionuclide produced by a reactor or cyclotron. A list of these radionuclides, together with their use and mode of production, is given in the Appendix.

The production of radionuclides requires the following stages of manufacture; preparation of a suitable target system, irradiation of the target and separation of the product radionuclide as a radiochemical from the target material. For *in vivo* use, the radiochemical is made into a safe and efficacious *radiopharmaceutical*. This requires conversion of the product to the desired chemical form and assurance of the physical, chemical and pharmaceutical qualities of the radiopharmaceutical.

The irradiation of targets in nuclear reactors and medical cyclotrons, the separation of the product radionuclide and its quality assurance, are described in more detail in the following sections.

1.2 Reactor Production of Radionuclides

Nuclear reactors provide a reliable source of neutrons for radionuclide production. In the reactor, intense neutron irradiation of the target element results in the addition of neutrons to the target nuclei and the formation of 'neutron-rich' radionuclides, in which the number of neutrons in the nucleus exceeds the number in the stable isotope of the same element. Baker (1966) has described the general aspects of production.

1

1.2.1   Production reactions

Radionuclides can be produced in a number of ways. The most common is the (n, γ) reaction using thermal neutrons accompanied by the emission of gamma radiation, e.g.

$$^{58}_{26}\text{Fe} + ^{1}_{0}\text{n} \rightarrow ^{59}_{26}\text{Fe} + \gamma \qquad \text{i.e.} \qquad ^{58}\text{Fe}(\text{n}, \gamma)^{59}\text{Fe}.$$

About two thirds of reactor-produced radionuclides are made in this way. Since the product is another isotope of the target element, it cannot be separated from the target by chemical means. This limits the specific activity which can be achieved.

At higher neutron energies (n, p) and (n, α) reactions can proceed. A small number of radionuclides are made by the former reaction, e.g.

$$^{58}_{28}\text{Ni} + ^{1}_{0}\text{n} \rightarrow ^{58}_{27}\text{Co} + ^{1}_{1}\text{p}$$

$$^{32}_{16}\text{S} + ^{1}_{0}\text{n} \rightarrow ^{32}_{15}\text{P} + ^{1}_{1}\text{p}.$$

Since the target element is transmuted to a different product element in this process, the product can be separated completely by chemical means. This results in products of high specific activity, limited only by the impurity levels in the processing reagents.

Occasionally the product of an (n, γ) reaction will decay rapidly, transforming to a radionuclide of a different element. This enables the latter to be separated chemically with high specific activity. An important example of this process is the production of iodine-131:

$$^{130}_{52}\text{Te} + ^{1}_{0}\text{n} \rightarrow ^{131}_{52}\text{Te} + \gamma \xrightarrow[25\,\text{min}]{\text{e}^-} {}^{131}_{53}\text{I}.$$

A number of radionuclides can be prepared as fission products of uranium-235 or plutonium-239. (Interaction of fast neutrons with ^{235}U results in the formation of plutonium-239.) Fission occurs as a result of the capture of a thermal neutron and the nuclei of elements lying near the centre of the periodic table are formed, e.g.

$$\text{U}(\text{n}, \text{f}) \rightarrow ^{131}_{52}\text{Te} \xrightarrow[25\,\text{min}]{\text{e}^-} {}^{131}_{53}\text{I}.$$

Since the fission products differ elementally from the original material, these radionuclides can be separated chemically with high specific activity. Such products are a natural consequence of reactor operation and may be isolated when spent fuel elements are reprocessed. However, because of the very high activities and the many elements present, this is a complex and hazardous operation. The more usual practice for commercial production is to irradiate a uranium-235 target. This method is employed for the production of ^{131}I, ^{133}Xe, ^{132}Te and ^{99}Mo. The last two radionuclides are important as the parent nuclides in generator systems (see § 1.5).

1.2.2 Irradiation

In practice the material for irradiation is placed in a sealed container and transported mechanically to and from the core of the reactor. Silica and aluminium containers are commonly employed. The following considerations should be observed in the choice of the target material.

(a) It should not endanger the operation of the reactor. For radiological and reactor safety upper limits have to be placed on the quantities of material with high neutron cross-sections.

(b) It should be in a suitable chemical form for subsequent processing to the desired compound. Few radiopharmaceuticals are immediate products of irradiation.

(c) It should be free from impurities in case these have high neutron capture cross-sections and give rise to radiocontaminants. It should also be free from impurities which are isotopic with the product radionuclide since they will reduce its specific activity.

1.2.3 Product yield

A number of processes control the yield and the maximum specific activity for the production of any individual radionuclide in a particular reactor. These are:

(a) The rate of production of nuclei of the desired radionuclide.

(b) The rate of destruction of the product nuclei by their own radioactive decay or by further neutron irradiation to form additional products (termed 'secondary neutron capture' or 'secondary burn-up').

(c) The rate of utilisation of the target nuclei.

Combination of these processes leads to the following expression for the yield of a product radionuclide;

$$A = \frac{N\varphi\sigma_1\lambda}{\lambda + \sigma_2\varphi - \sigma_1\varphi} \{\exp(-\sigma_1\varphi t) - \exp[-(\lambda + \sigma_2\varphi)t]\} \quad (1.1)$$

where A is the radioactivity, N is the initial number of target nuclei, σ_1 is the cross-section of the target nuclei, σ_2 is the cross-section of the product nuclei, λ is the decay constant of the product nuclei, φ is the neutron flux and t is the irradiation time. Enumeration of this expression shows that the radioactivity rises to a maximum and then falls due to target conversion. Often the expression can be simplified because target and secondary burn-up are negligible (the latter only becomes important in high flux or long irradiations). It can then be stated as:

$$S = \frac{0.6\sigma\varphi(1 - e^{-\lambda t})}{W} \qquad (1.2)$$

where S is the specific activity of the product (Bq g^{-1}), φ is the effective neutron flux in the sample (neutrons cm^{-2} s^{-1}), σ is the activation cross-section of the target nuclei (barns), W is the atomic weight of the target nuclei (g), t is the irradiation time and λ is the decay constant of the product. This is a standard 'inverse exponential' or 'build-up' expression. At large values of t, S reaches a saturation value given by:

$$S = \frac{0.6\sigma\varphi}{W}. \qquad (1.3)$$

Half the saturation activity is reached after irradiation for one half-life and little activity is gained by irradiation beyond three half-lives.

The key factors in the production rate are the neutron flux and the cross-section of the target nuclei. The highest possible flux density should always be used. The yield can be increased by increasing the proportion of target nuclei in the target mass using 'enriched' or 'separated' target materials. Electromagnetic enrichment using mass spectrometry is usually employed; Rupp (1973) lists 232 nuclides which have been enriched at the Oak Ridge National Laboratory, USA. Enrichment has other advantages besides increasing the yield. Firstly, enrichment of one nuclide requires that other nuclides must be depleted and this results in a purer product with fewer radiocontaminants than irradiation of the natural material. For example, neutron irradiation of natural iron (5.82% ^{54}Fe, 91.66% ^{56}Fe, 2.19% ^{57}Fe and 0.33% ^{58}Fe) results in 90% ^{59}Fe with 10% ^{55}Fe. Enriching the ^{58}Fe content of the target to 85% results in less than 5% ^{55}Fe. Enriching the target to 98% ^{54}Fe yields 100% ^{55}Fe. Secondly, where the product has a short half-life, a higher yield can be obtained before equilibrium is reached.

In practical circumstances there is always some uncertainty about the yield expected. This is due to uncertainties in the effective neutron flux and the neutron energy spectrum at the irradiation site. These are caused by non-uniformities introduced by the target and its handling equipment.

The dependence of yield on the energy spectrum is particularly critical because interaction cross-sections alter markedly with neutron energy.

Fuller details of targets, irradiation techniques and product processing for 15 common reactor-produced radionuclides are given in the Manual of Radioisotope Production (IAEA 1966). A less detailed, but more comprehensive, list of production reactions is given in the Radiochemical Manual (1966).

1.3 Cyclotron Production of Radionuclides

In the medical cyclotron bombardment of the target element by energetic charged particles, such as protons, deuterons and α-particles, results in the addition of these particles to the target nuclei and the formation of 'neutron deficient' radionuclides.

Routine production of radionuclides using a cyclotron is more complex and more costly than in the reactor. This is due to the cyclotron's less stable operating characteristics, greater maintenance requirements and more complex targets. However problems of unreliability are rapidly being overcome and cyclotrons have become versatile producers of a range of very useful radionuclides in the past few years.

Production takes place in cyclotrons for a number of reasons:

(a) Charged particle bombardment may be the only means of production of a particular radionuclide. This applies to ^{11}C, ^{13}N and ^{15}O which are the only γ-ray emitting radionuclides of the biologically important elements.

(b) The cyclotron provides a higher specific activity product than that available by neutron irradiation. Charged particle bombardment always results in the transmutation of the target nuclei and enables a high specific activity to be achieved by chemical separation. For example, the production of ^{51}Cr by the $^{51}V(d, 2n)^{51}Cr$ reaction instead of the $^{50}Cr(n, \gamma)^{51}Cr$ reaction.

(c) A radionuclide produced by a cyclotron may have more favourable radiation characteristics as regards detection and/or radiation exposure than reactor-produced radionuclides of the same element. For example, the sites of red cell production may be imaged safely with cyclotron-produced ^{52}Fe but not with the more widely available reactor-produced ^{59}Fe, because the latter has poorer detection characteristics and delivers a higher radiation dose per unit activity. ^{59}Fe may only be administered safely in small quantities inadequate for imaging purposes.

Reasons for the adoption of the cyclotron for radionuclide production, in preference to other charged particle accelerators, are given by White-house and Putnam (1953). In brief, Van de Graaff generators cannot accelerate particles sufficiently fast and betatrons and electron synchro-trons only accelerate electrons, which produce x-radiation when stopped by a target. The more energetic particles available in the synchro-cyclotron cause the disintegration of the target nuclei with the formation of a complex mixture of products. Recently, proton linear accelerators, originally built as injectors for accelerators for high energy physics research, have been used for the production of radionuclides using spallation reactions, e.g. the Brookhaven Linac Isotope Producer (BLIP) (Richards *et al* 1973).

1.3.1   Production reactions

Four types of charged particle are employed for radionuclide production in cyclotrons. These are protons, deuterons, $^3He^{2+}$ and α-particles. In small medical cyclotrons these are accelerated to energies of up to 40 MeV, 20 MeV, 90 MeV and 70 MeV respectively.

The majority of radionuclides are produced by deuteron or α-particle bombardment. The (d, n) reaction is commonly used to produce the short-lived radionuclides of the biologically important elements, carbon, nitrogen and oxygen, i.e. $^{10}B(d, n)^{11}C$, $^{12}C(d, n)^{13}N$ and $^{14}N(d, n)^{15}O$. The optimum deuteron energies for these reactions are 15, 15 and 6.5 MeV. A number of reactions may occur with α-particle bombard-ment, e.g. (α, pn) and (α, 2n). Examples of these reactions for the production of medically useful radionuclides are $^{16}O(\alpha, pn)^{18}F$ and $^{121}Sb(\alpha, 2n)^{123}I$.

Bombardment with protons or $^3He^{2+}$ particles is less common. Exam-ples which have been exploited practically are $^{14}N(p, \alpha)^{11}C$, $^{124}Te(p, 2n)^{123}I$ and $^{16}O(^3He, p)^{18}F$. The first reaction used a proton energy of 7.4 MeV, the second a proton energy of 14 MeV and the last a $^3He^{2+}$ energy of 30 MeV.

In larger cyclotrons, beams of higher energy protons have been used to produce radionuclides by spallation reactions. Two recent practical examples are:

$$^{127}I\,(p, 5n)\,^{123}Xe \xrightarrow{e^-} {}^{123}I$$

and

$$^{133}Cs\,(p, 2p\,5n)^{127}Xe.$$

Silvester (1973) comprehensively reviewed the accelerator production of

46 medically useful radionuclides and has listed their production reactions. More recent information on medical cyclotrons is given by Serafini and Beaver (1978).

1.3.2 Target design

Internal and external target configurations are used in cyclotrons. Internal targets are sited in the dee chamber towards its edge and intercept the internal beam. Most of the energy in the high current internal beam is converted to heat and this has to be dissipated to prevent evaporation of the target material. For example, in the MRC cyclotron at Hammersmith Hospital, London, the internal beam at 15 MeV with a current of 500 μA and a cross-section $10 \times 3 \, mm^2$ has a power density of $20 \, kW \, cm^{-2}$ (Vonberg *et al* 1970). Water cooling at high pressure is usually employed but this is insufficient if the target is placed at normal incidence to the beam. To reduce the power density at the target surface, the target is either placed at grazing incidence to the beam, or the beam and target are moved relative to one another in a regular way. The most common practice in the latter case is to use a rotating target, but this is mechanically complex and expensive.

Internal targets for high beam currents can be prepared in a number of ways but are clearly limited to materials with suitable physical properties. Flat plate targets are prepared by depositing the target material onto a flat copper or aluminium plate. Bonding is optimised by using the most appropriate technique, e.g. electroplating, metal spraying, rolling, etc. Capsule targets are used for the irradiation of metal powders or high melting point chemical compounds, usually oxides. A one third loss of beam current will occur at the leading edge of the capsule. In both cases the target backing plate and the capsule are in intimate contact with the cooling water.

External targets are sited outside the dee chamber. The beam is extracted from the chamber and shaped and guided to the target by magnets. The beam can also be defocused by magnets to cover a larger area, with a consequent reduction in power density. This enables targets to be placed normal to the beam. Much lower beam currents are available in this configuration, e.g. the internal beam of 500 μA of the MRC cyclotron is reduced to 100 μA (Vonberg *et al* 1970). Foil windows are usually placed in the beam line to maintain the vacuum of the dee chamber, and over the target to contain the target material and bombardment products. This arrangement, called the window target, enables solid, liquid and gaseous targets to be handled.

A more detailed consideration of target design and operation has been given by Vonberg *et al* (1970), who have also listed the major factors in target design for efficient radionuclide production. These are:

(a) The physical properties and chemical purity of the target material.
(b) The provision of adequate cooling to control the temperature rise of the target surface.
(c) The method of recovery of the product during or after bombardment.
(d) The selection of beam energy and target thickness to maximise the yield of the product and minimise the yield of possible contaminants. External targets are better in this respect since they readily allow variations in target thickness, whereas an internal target at grazing incidence is always a thick target (in the nuclear sense).

External targets are of course more convenient to change at the end of a bombardment than internal targets and with a lower radiation exposure for the staff involved. Multiple beam lines to a number of external targets may also be employed.

1.3.3 Product yield

Svoboda (1970) has examined the calculation of the yield of radionuclides by charged particle bombardment. Expression (1.1) describing the yield of a nuclear reaction is applicable to reactor production since the energy of the bombardment particles (usually neutrons), and hence the reaction cross-section, does not change as the particles penetrate into the target.

It is also applicable to the charged particle bombardment of thin targets (in the nuclear sense), where the cross-section does not change appreciably with depth. Equation (1.2) can then be more conveniently stated as

$$A = K\sigma \frac{I}{Z} \frac{S}{N} (1 - e^{-\lambda t}) \qquad (1.4)$$

where A is the radioactivity produced, σ is the cross-section of the target nuclei, I is the beam current, Z is the number of elementary charges on the bombarding particles, S is the target thickness (mass per unit area), N is the number of target nuclei and K is a constant. In practice, S must be less than 10 mg cm^{-2}.

However expressions (1.1) and (1.2) cannot be applied to most cyclotron production, where a thick target is generally used, because the energy of the bombarding particles decreases rapidly on passing into the target which causes large changes in the reaction cross-section with depth of penetration. Consequently the formula must be more complex. In addition, if the target material is not contained within a capsule or behind a window, it may be necessary to take account of evaporation of the product due to overheating of the target.

Two further factors limit the accuracy of calculating the yield. The first is that the data on cross-sections for charged particle reactions (termed 'excitation functions') are limited. The second is that it is difficult to measure the energy and the intensity of the bombarding particles.

In the face of these theoretical uncertainties a convenient practical term, the 'thick target yield', is widely used. This is supposed to be the radioactivity produced in an irradiation lasting one hour. This yield Y (kBq $\mu A^{-1} h^{-1}$ or MBq $mA^{-1} h^{-1}$) is given by:

$$Y = A/It \tag{1.5}$$

where A is the activity (kBq or MBq), I is the bombarding particle beam intensity (μA or mA) and t is the irradiation time (h). This definition implies that yield is proportional to irradiation time, but clearly the relative lengths of irradiation time and the half-life of the product must be taken into account. Provided the irradiation time is less than one tenth of the half-life, the error in yield using equation (1.5) will not exceed 4%.

As an alternative, Svoboda (1970) has proposed the use of the production rate R, or 'saturation activity' of a thick target for 1 μA. The activity, A, of a product is then given by:

$$A = RI(1 - e^{-\lambda t}). \tag{1.6}$$

When the irradiation time, t, is short by comparison with the product half-life, then:

$$A \approx RI\lambda t \qquad \text{i.e.} \qquad Y = \lambda R.$$

The arguments in favour of the use of R are:

(a) it is independent of irradiation time,
(b) it is independent of the half-life of the product, and
(c) it is readily evaluated from known cross-sections and a given energy.

Svoboda proposes that R should always be used for the production of radionuclides with half-lives less than one day. No agreement has however been reached on its use, though it is often used in activation analysis. It has been used by Syme *et al* (1978) to calculate the yields for the UK production of ^{123}I on the Harwell variable energy cyclotron. This paper also demonstrates the difficulties of measuring excitation functions accurately.

1.4   Product Separation

After the irradiation of solid targets, the product radionuclide has to be separated chemically from the residual target material. This may be by simple dissolution of the product from the target or be more complex and employ ion exchange, precipitation or distillation. The method of recovery is an important feature in the selection of the target material and should be the simplest available. Dissolution is the most common method; for example water is employed in the recovery of ^{82}Br, ^{42}K and ^{24}Na and concentrated hydrochloric acid in the recovery of ^{58}Co and ^{59}Fe. Distillation is employed for the separation of ^{32}P. Other examples can be found in the Manual of Radioisotope Production (IAEA 1966). The recovery process can also concentrate the product and remove impurities.

Continuous recovery of the product is often practicable with gaseous and liquid targets in cyclotrons by the continuous circulation of the target material through the target and separation equipment. Gaseous examples are the production of ^{15}O from nitrogen using the ^{14}N(d, n)^{15}O reaction and ^{43}K from argon using the ^{40}A(α, p)^{43}K reaction. An example of a liquid target is di-iodomethane ($C_2H_2I_2$), used in the production of ^{123}I on the Harwell variable energy cyclotron (Cunninghame *et al* 1978).

1.5   Radionuclide Generators

The radionuclide generator is a convenient means of providing a continuous supply of a short-lived nuclide at sites distant from a reactor or cyclotron. The physical basis of generator operation is a parent–daughter decay scheme in which the daughter can be readily and repeatedly separated from the parent. The daughter is used in clinical investigation and its short half-life results in low patient radiation dosage. The parent has a relatively long half-life to give the generator a working life of reasonable length.

The first application of the generator principle was the development of the ^{132}Te–^{132}I generator by the Brookhaven National Laboratory, USA in 1954. The second development was the ^{99}Mo–^{99m}Tc (^{99}Tc in a metastable state) generator in 1967, again by the Brookhaven National Laboratory. Richards (1966a) listed a number of existing and possible generator systems. The generators commonly used in nuclear medicine today are listed in table 1.1, together with the half-lives of the parent and daughter radionuclides.

The ^{99}Mo–^{99m}Tc generator has become the most common generator, being found in almost all hospitals for the provision of ^{99m}Tc for organ visualisation agents. The ^{113}Sn–^{113}In generator is used to a much lesser extent for the same purpose. The ^{87}Y–^{87m}Sr and ^{132}Te–^{132}I generators are becoming less common as the daughter nuclides are superseded by others with more favourable characteristics.

Table 1.1   Radionuclide generators.

Parent		Daughter		
Element	Half-life	Element	Half-life	Product
^{81}Rb	4.5 h	^{81m}Kr	13.5 s	^{81}Kr†
^{87}Y	80 h	^{87m}Sr	2.8 h	^{87}Sr
^{99}Mo	67 h	^{99m}Tc	6 h	^{99}Tc†
^{113}Sn	118 h	^{113m}In	1.7 h	^{113}In
^{132}Te	78 h	^{132}I	2.3 h	^{132}Xe

† Half-lives greater than 10^5 years.

1.5.1   Physical properties

The general relationships of parent–daughter schemes are well known and are based on the decay constants of the nuclides.

The activity A_1 of the parent is given by the radioactive decay equation:

$$A_1 = A_1^0 e^{-\lambda_1 t} \tag{1.7}$$

where A_1^0 is the initial activity of the parent and λ_1 its transformation constant. The initial activity of a generator is specified at a given date and time and a range of activities is available from commercial suppliers.

The activity of the daughter nuclide, A_2, will depend on (a) the rate of decay of the parent which equals the formation rate of the daughter,

(b) the rate of decay of the daughter (transformation constant λ_2) and
(c) the time since the last separation of the daughter from the parent
(t). The activity of the daughter is given by:

$$A_2 = \frac{\lambda_2}{\lambda_2 - \lambda_1} A_1^0 (e^{-\lambda_1 t} - e^{-\lambda_2 t}). \qquad (1.8)$$

If the daughter has a much shorter half-life than the parent
($\lambda_1 \ll \lambda_2$) as in the ^{113}Sn–^{113m}In generator, the decay of the parent can
be ignored ($e^{-\lambda_1 t} = 1$). Equation (1.8) then becomes:

$$A_2 = A_1^0 (1 - e^{-\lambda_2 t}).$$

When t is large, $A_2 = A_1^0$ and the activity of the daughter equals that of
the parent.

In the general case described by equation (1.8), the activity of the
daughter builds up after separation and then falls, in parallel with the
activity of the parent. When t is sufficiently large (about five half-lives
of the daughter), $e^{-\lambda_2 t}$ is small by comparison with $e^{-\lambda_1 t}$, transient equilibrium occurs and equation (1.8) becomes:

$$A_2 = \frac{\lambda_2}{\lambda_2 - \lambda_1} A_1^0 e^{-\lambda_1 t}. \qquad (1.9)$$

The time of maximum daughter activity after separation is given by:

$$t = \frac{1}{\lambda_2 - \lambda_1} \ln \frac{\lambda_2}{\lambda_1}. \qquad (1.10)$$

This situation applies to the ^{99}Mo–^{99m}Tc generator.

In the ^{81}Rb–^{81m}Kr generator continuous elution is employed by passing
air or a saline solution over the parent. In this case the number of atoms
of the daughter produced per unit time will equal the activity of the
parent (A_1). The activity of the daughter will be given by $A_2 A_1$.

1.5.2   Practical considerations

For regular use, a generator must be simple and convenient and rapid
to operate. In all commercially produced generators, the daughter is a
different element from the parent and a difference in chemical behaviour
is the basis of separation. Physical techniques can also be employed,
namely distillation, sublimation or solvent extraction, but these are more
complex to perform and require more elaborate equipment. Of the
procedures available for chemical separation, the simplicity of the ion

exchange mechanism makes it the method of choice. This is the basis of the chromatographic generator.

During manufacture of the chromatographic generator, the parent nuclide is absorbed onto a column of ion exchange material. The column may be an organic material, e.g. resin, or an inorganic absorbent, such as aluminium oxide or zirconium oxide. Inorganic absorbers are preferred because they have a greater resistance to radiation damage and are less susceptible to contamination by pyrogenic materials (see §1.6.5). A typical column contains 5 g of material and is contained within a glass or plastic tube closed at each end by scintered glass caps. The whole is contained in a lead shield to provide radiological protection. Daughter activity is eluted from the generator by passing the appropriate reagent, the eluant, through the column and collecting the product solution, the eluate.

The practical features of generator design and operation are best illustrated by reference to the commonest and most developed generator—the ^{99}Mo–^{99m}Tc generator. The parent ^{99}Mo is prepared as ammonium molybdate (NH_4MoO_4) and absorbed onto an aluminium column. The ^{99}Mo decays and ^{99m}Tc is formed as the pertechnetate ion TcO_4^-. When a saline solution is passed through the column, the chloride ions exchange with the pertechnetate ions but not with the molybdate ions. The solution leaving the column contains sodium pertechnetate ($NaTcO_4$). Richards (1966b) has described the early form of this generator. Since then its detailed design has been altered to simplify operation and to assure the quality of the eluate. A modern ^{99}Mo–^{99m}Tc generator made by The Radiochemical Centre is shown in figure 1.1. In this generator, elution is made automatic by putting a pressurised vial of saline solution on the input line of the generator. The air pressure in the vial forces the saline solution through the column and into the collection vessel which has an air bleed (positive pressure generator). Automatic elution is achieved in other manufacturers' generators by putting an evacuated vial onto the outlet of the generator and drawing saline solution through the column from a reservoir (negative pressure generator).

The parent ^{99}Mo may be prepared either by fission of ^{235}U or by the irradiation of enriched ^{98}Mo in a reactor. The former is a more efficient process and has the advantage of producing carrier-free ^{99}Mo. Saturation irradiation of 100 mg of ^{235}U in a flux of 10^{13} neutrons cm^{-2} s^{-1} yields approximately 2.2 Ci of ^{99}Mo; the same irradiation of 1 g of molybdenum will yield about 200 mCi. Since the fission product is carrier-free, larger activity may be adsorbed onto the column of the generator without danger of elution of parent activity. In practice the column size may also be reduced and a smaller volume of saline is necessary to completely elute the daughter activity. Typically 15 ml of saline solution is required to elute an enriched ^{99}Mo generator and 5 ml for a fission ^{99}Mo generator. This is an advantage when the activity is required in a small volume, e.g. for blood flow studies.

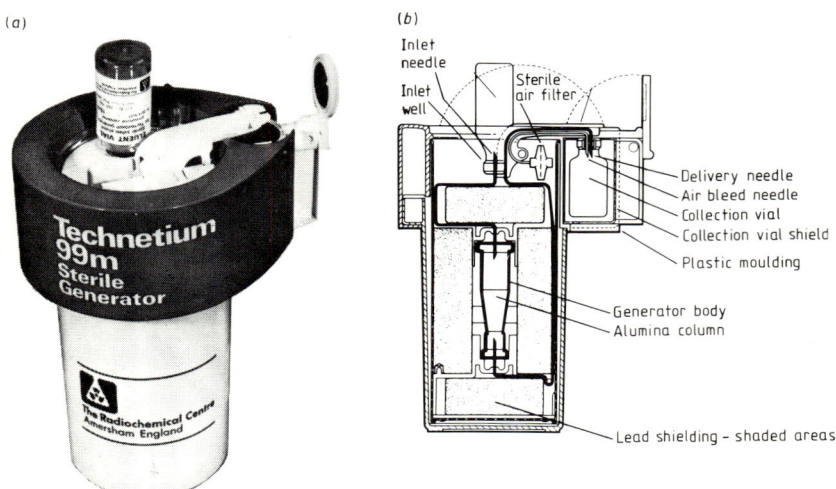

Figure 1.1   Radionuclide generator. (*a*) A positive pressure ^{99m}Tc generator with (*b*) sectional elevation. Saline solution passes from the up-ended pressurised vial through the alumina column to the collection vial. (Reproduced by kind permission of The Radiochemical Centre Ltd.)

Boyd (1973) has examined the developments in ^{99}Mo–^{99m}Tc generators and in particular the parameters of performance. He cites four such parameters, namely the elution efficiency of the daughter, the elution profile, the radionuclide purity and the chemical purity. The elution efficiency is the yield of ^{99m}Tc as a fraction of the instantaneous ^{99}Mo activity; this is typically 80% for an enriched ^{99}Mo generator and 90% for a fission ^{99}Mo generator. The elution efficiency can be severely reduced when there is radiolysis of the pertechnetate formed on the column. This results in the formation of lower valency species of ^{99m}Tc which are strongly absorbed by the alumina and retained during elution.

The elution profile is the yield of ^{99m}Tc against the volume of eluant passed through the generator. Typical profiles for the two types of generator are shown in figure 1.2. The narrower profile of the fission ^{99}Mo generator is due to the smaller column size and indicates the higher concentration of ^{99m}Tc available. The eluate should only contain one radionuclide ^{99m}Tc. The radionuclide purity will be reduced by the presence of the parent (^{99}Mo break-through) or of other contaminants resulting from target preparation and irradiation. These will increase the radiation dose to the patient. Good purity is achieved by reliable chemical processing, generator assembly and testing. The chemical purity of the eluate may be reduced by the presence of the column material, in this instance soluble and insoluble aluminium, both of which may adversely affect further chemical manipulation of the sodium pertechnetate solution. Soluble aluminium is removed by thorough washing of the column during assembly. Insoluble aluminium would appear to be due to radiation damage of the column and is removed by a membrane filter in the generator outlet.

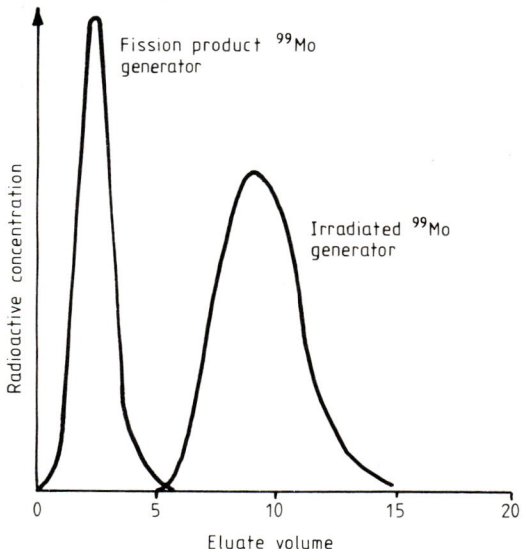

Figure 1.2 Elution profiles for ^{99m}Tc generators. The fission product ^{99}Mo generator having a smaller column elutes with a smaller total volume.

1.6   Quality Assurance

Radioactive compounds for *in vivo* use are termed 'radiopharmaceuticals'. Unlike conventional pharmaceuticals, they do not have a
pharmacological effect because of the small quantity of material administered. Most are used for investigative purposes and are administered
intravenously. A few are used for therapy through internal irradiation
(see Appendix).

Radiopharmaceuticals must meet stringent radioactive and pharmaceutical standards for safe and efficacious use. The quality of a
radiopharmaceutical will depend on the following characteristics:
radioactivity, radionuclidic purity, chemical purity, tissue specificity,
sterility, freedom from pyrogens and from foreign particulate matter.
These characteristics are considered separately in detail below. High
standards of the first four characteristics are necessary for radiopharmaceuticals administered by all methods; the latter three characteristics
are only important for injections.

1.6.1   Radioactivity

Measurement of the radioactivity of a dose of a radiopharmaceutical is
an essential and final part of its preparation. Allowance for radioactive
decay has to be made for doses to be administered later. The activity
of the dose will be the minimum acceptable for accurate results.

Preparation of doses is usually done on a volumetric basis using
activity concentration values supplied by the manufacturer, or measured
concentrations in the case of generator eluates. Checking of the dose
activity of radionuclides emitting γ-rays and energetic β-particles (e.g.
^{32}P) is done directly with a well type air ionisation chamber and an
electrometer (see §4.2). Checking of doses of low energy β-emitting
radionuclides, e.g. ^{3}H, is usually done by assaying an aliquot of the dose
in a liquid scintillation counter (see §4.1). Typically an accuracy of
±5% is accepted.

During investigations the radioactivity in samples and in regions of
the body is not expressed in curies or becquerels due to the difficulty
of making absolute measurements. Instead activities are expressed more
accurately as fractions or percentages of the administered dose. For this
purpose a known aliquot of the dose is retained and its activity determined under similar conditions to that of the sample or region of the
body.

1.6.2 Radionuclidic purity

Radionuclidic purity is defined as the percentage of the total radioactivity present as the stated radionuclide. This is not a constant value, but will depend on the relative half-lives of the stated nuclide and any contaminant nuclides. Contaminants with longer half-lives than the stated nuclides are potentially more hazardous because they will increase the radiation dose to the patient. They must therefore be reduced to an acceptable level at the time of production. An example which occurs during the production of ^{58}Co by the ^{58}Ni(n, p)^{58}Co reaction is the formation of ^{60}Co by the ^{59}Ni(n, γ)^{60}Co reaction, due to the presence of ^{59}Ni in the target. ^{60}Co has a half-life of 5.3 yr and a more damaging radiation spectrum than ^{58}Co, which has a half-life of 71 d. Less than 1% of the total activity must be due to ^{60}Co on the expiry date of the preparation. This example and the example in §1.2.3 of enriched targets for the production of ^{59}Fe or ^{55}Fe demonstrate the importance of target selection and purity on the purity of the final product. In a few instances the production process may have to take account of contaminant formation. For example, in the production of ^{198}Au by the ^{197}Au(n, γ) ^{198}Au reaction, the interaction cross-section of the product ^{198}Au cannot be ignored. This has a capture cross-section for thermal neutrons 268 times that of ^{197}Au, and ^{199}Au is readily formed by the ^{198}Au(n, γ)^{199}Au reaction. The ^{197}Au(n, γ)^{198}Au reaction has a peak cross-section at a neutron energy of 4 eV. The proportion of ^{199}Au is minimised either by using a short production time in comparison with the half-life of ^{198}Au, or by enclosing the target in cadmium foil. The latter absorbs the thermal neutrons and prevents the secondary reaction. Not more than 5% of the activity present may be due to ^{199}Au.

Contaminants with a shorter half-life than the stated nuclide are less of a problem because storage will reduce the contaminant to a level acceptable for clinical use. One example is the presence of ^{126}I in ^{125}I produced by the reaction ^{124}Xe(n, γ)^{125}Xe $\xrightarrow{e^-}$ ^{125}I due to secondary neutron capture by the product ^{125}Xe and subsequent decay. ^{126}I and ^{125}I have half-lives of 13 d and 60 d respectively and preparations are released when the former constitutes 1% or less of the total activity. A second example is ^{66}Ga contamination of ^{67}Ga produced by the ^{65}Cu(α, 2n)^{67}Ga reaction. The target is natural copper which also contains ^{63}Cu; the contaminant is produced by the ^{63}Cu(α, n)^{66}Ga reaction. ^{66}Ga and ^{67}Ga have half-lives of 9.3 h and 78 h respectively and the ^{66}Ga content falls to an acceptable level after one to two days.

In radionuclides produced by generators, the most probable contaminant is the parent nuclide. The generator manufacturer has to ensure that the level of the parent nuclide is below the acceptable limit in normal usage throughout the generator's working life. The acceptable levels are 0.1% of the ^{99m}Tc activity for ^{99}Mo and 0.01% of the ^{113m}In activity for ^{113}Sn. Other longer-lived contaminants may be present as a result of impurities in the production of the parent. These have included ^{60}Co, ^{86}Rb, ^{110m}Ag, ^{131}I, ^{131}Ba, ^{134}Cs and ^{239}Ne (Müller and Steinnes 1971); for the ^{99m}Tc generator these contaminants should not exceed 0.01% of the total activity.

Checking for γ-ray emitting impurities can be performed by examining the radiation spectrum for radiations other than those of the stated nuclide. This is done using a shielded solid state detector with good energy resolution, e.g. a lithium drifted germanium detector and multichannel analyser. Detection of small quantities of impurities may be difficult in the presence of large activities of the stated nuclide and the sample may have to be chemically separated or allowed to decay before spectral analysis can proceed.

1.6.3　Chemical purity and specific activity

Chemical purity is the fraction of the total mass present in the stated chemical form. Chemical impurities arise from the production process. In generator produced radionuclides, the most probable contaminant is the column material, e.g. aluminium.

Since most radiopharmaceuticals are produced with high specific activity, the mass present is very small, e.g. the mass of 400 MBq of sodium (^{99m}Tc) pertechnetate is only 2 ng. Even low levels of impurities can therefore interfere with the desired chemical reactions and levels must be kept extremely low in such applications.

There are two reasons for using a high specific activity material in order to reduce the mass of stable material administered. Firstly, the material may be toxic, e.g. chromium. Sodium (^{51}Cr) chromate used to label red blood cells (see §5.3) should have a specific activity greater than 800 MBq mg^{-1} on the reference day. A typical dose of 4 MBq ^{51}Cr results in the administration of less than 5 µg of sodium chromate. Secondly, when investigating physiological or pathological processes it is important to administer an amount of material which is small enough not to perturb the process under investigation, i.e. so that the administered radioactivity has true trace and tracer roles. Examples are ferric (^{59}Fe) citrate for ferrokinetic investigations (see §6.3.2) (specific activity

> 40 MBq mg^{-1}) and sodium (^{131}I) iodide for measurement of thyroid function (see §6.4.3) (specific activity > 200 MBq µg^{-1}).

1.6.4 Tissue specificity

The characteristic which ensures that the radiopharmaceutical reaches a target organ or follows a desired metabolic pathway is most often its chemical state, and less frequently its physical state.

The chemical properties which determine the behaviour of a radio-pharmaceutical *in vivo* are the radiochemical purity and the pH.

The radiochemical purity is defined as the percentage of the total radioactivity present in the stated chemical form. The presence of radiochemical impurities, i.e. radioactivity in other chemical forms, in a radiopharmaceutical will generally have two adverse effects. Firstly, the results of the investigation may be misleading and of poorer quality because of interference from impurities with different physiological behaviour. Secondly, non-target organs may be unnecessarily irradiated. For example, free iodide (^{131}I, ^{125}I or ^{123}I) or pertechnetate (^{99m}TcO$_4$) in more complex radiopharmaceuticals with these labels will always go to the thyroid gland unless gland uptake is prevented.

The presence of radiochemical impurities can be detected by suitable chemical separation techniques; chromatography is most commonly employed. After separation, the quantities of the different components present are measured by determining their separate activities using a radiochromatogram scanner (see §4.3).

pH is critical in only a few radiopharmaceuticals e.g. indium (^{113m}In) chloride for labelling transferrin *in vivo*. For most radiopharmaceuticals, pH is not critical and is adjusted to the range 6.5–8.5 for compatibility with plasma pH.

The only physical property which is used to determine the behaviour of radiopharmaceuticals is particle size. Radioactive particles with diameters in the range 15–50 µm are used to demonstrate tissue perfusion by virtue of their trapping in blood capillaries. Colloidal particles with diameters in the range 200–800 nm are used to demonstrate the function of the reticulo-endothelial system through their phagocytic uptake.

1.6.5 Further standards for injections

Radiopharmaceuticals administered by injection must, in addition to the four characteristics above, be sterile, pyrogen-free and free from foreign particulate matter.

Sterility is the absence of living organisms, usually bacteria and spores. This can be approached in one of three ways: (a) by terminal sterilisation of the dose in its final container, e.g. autoclaving, (b) passage of the dose through a bacteria proof filter (with a porosity of 0.45 or 0.22 μm) into the final container, or (c) by aseptic methods starting from sterile materials. The effectiveness of the sterilisation process decreases from method (a) to method (b) to method (c).

Pyrogens are the metabolic products of bacteria, yeasts and fungi and can therefore be present as a result of present or past microbiological contamination. Sterility is not therefore an indication of freedom from pyrogen and none of the above methods of sterilisation destroy pyrogens. When given in an injection, pyrogens can give rise to fever and joint pains. Freedom from pyrogens is achieved by using pyrogen-free containers, sterile materials and aseptic procedures throughout the production process.

The presence of particles in injections can lead to the formation of micro-emboli and granulomata in the body. Freedom from particles can be achieved with a clean production environment, effective cleaning procedures for glassware and particle-free ingredient solutions. The latter can be achieved both for reagents and the product by passage through a 1.2 μm porosity filter.

2 Radiation Effects

2.1 Introduction

The selection of a radionuclide in an appropriate chemical form to investigate some physiological process depends on two main factors. These are:

(a) The known behaviour of the material in the body.

(b) The radiation characteristics of the radionuclide which will influence the type of information that can be gained and the radiation dose to the patient.

The first of these factors has a physiological and biochemical basis and will not be discussed further here. The latter has a physical basis which is outlined below.

The transition in the nucleus from a high energy state to a lower energy state during radioactive decay is always accomplished by the emission of charged particles and/or γ-rays. In *in vivo* applications, the part of this radiation which emerges from the organism is, of course, the means by which a radionuclidic tracer is detected and quantified. The remainder, which is absorbed by the organism, will deposit energy giving rise to biological damage. Ideally the former should be maximal and the latter minimal. Both effects will vary in magnitude with radiation type and energy. The decay scheme of a radionuclide is therefore an important factor in deciding its suitability for biological use.

2.2 Decay Mechanisms

2.2.1 Alpha decay

Unstable nuclides with high atomic numbers have an excess of nucleons. They achieve stability through the emission of α-particles, e.g.

$$^{226}_{88}\text{Ra} \rightarrow {}^{222}_{86}\text{Rn} + {}^{4}_{2}\alpha.$$

2.2.2 Beta decay

At lower atomic numbers, unstable nuclides have a different neutron to proton ratio from that in the stable nuclide. Reactor-produced radio-nuclides have an excess of neutrons and achieve stability by the conversion of a neutron to a proton. This is accompanied by the emission of an electron and an anti-neutrino, ν^*, i.e.

$$n \rightarrow p + e^- + \nu^*.$$

An example of pure beta decay is the decay of carbon-14:

$$^{14}_{6}C \rightarrow ^{14}_{7}N + e^- + \nu^*.$$

The energy difference between the parent and daughter nuclides is shared between the electron and anti-neutrino with the amount taken by each varying from zero to the maximum energy. Typically the energy of the electron is one third the maximum energy.

2.2.3   Positron decay and electron capture

Accelerator-produced radionuclides have an excess of protons and achieve stability by the conversion of a proton to a neutron. A positron (e^+) and a neutrino ν are emitted during the process, i.e.

$$p \rightarrow n + e^+ + \nu.$$

As in beta decay the positrons have a range of energies up to a maximum energy, the balance at energies below the maximum being imparted to the neutrino. An example of positron decay is the decay of carbon-11:

$$^{11}_{6}C \rightarrow ^{11}_{5}B + e^+ + \nu.$$

The positron loses its energy by excitation and ionisation interactions along its path. When its energy is dissipated, it is annihilated by combination with an electron. The mass of the two particles is converted to electromagnetic radiation and two photons (annihilation radiation) are produced:

$$e^+ + e^- \rightarrow 2\gamma.$$

The photons each have an energy of 0.51 MeV and travel in diametrically opposite directions (to conserve the zero momentum of the system).

An alternative mechanism for achieving the same alteration in the neutron to proton ratio is the capture of an orbital electron by the nucleus. This electron is from the K shell in about 90% of decay schemes and from the L shell in the remainder. The electron combines with a proton to form a neutron with the emission of a neutrino, i.e.

$$p + e^- \rightarrow n + \nu.$$

An example of electron capture is the decay of caesium-131:

$$^{131}_{55}\text{Cs} + e^- \rightarrow ^{131}_{54}\text{Xe} + \nu.$$

The capture of an electron by the nucleus leaves a vacancy in one of the inner electron shells. As a result the product atom, and not the nucleus, has an energy above the ground state. This excess energy is lost either by the emission of the characteristic x-radiation of the daughter, or by the transfer of the excess energy to an outer electron, thereby ejecting it as an Auger electron. The x-rays and Auger electron have much lower energies than the particle emissions since the former result from extra-nuclear processes and the latter from intra-nuclear processes.

Electron capture is more likely to occur than positron decay for nuclides with high Z in which the K shell is closer to the nucleus. Positron decay will predominate at low Z provided there is an energy difference of at least 1.02 MeV between the parent and daughter nuclides.

2.2.4 Gamma radiation

The majority of radionuclides do not decay with simple schemes as described above. Often the charged particle radiation carries off only part of the energy difference between the parent and daughter. The product nucleus is then formed in an excited state and reaches the ground state by the emission of gamma radiation. The γ-rays emitted have discrete energies because they correspond to the energy difference between two characteristic energy states of the daughter after the decay process has taken place. In the majority of cases the transitions occur virtually instantaneously. Often there are several modes of decay, passing through different energy states of the daughter. For example, cobalt-60 decays by beta decay with the emission of two γ-ray energies according to the following scheme:

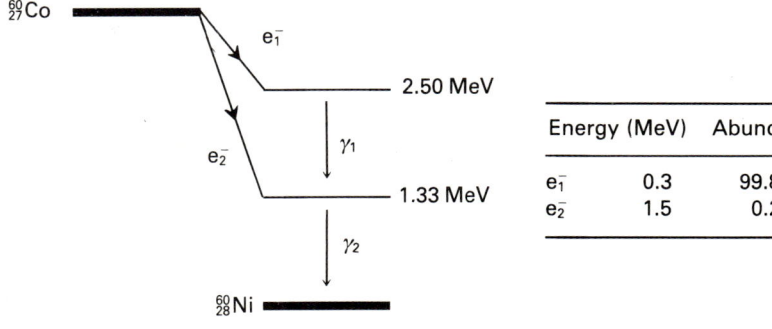

Energy (MeV)	Abundance
e_1^-	0.3	99.8%
e_2^-	1.5	0.2%

Chromium-51 decays by electron capture with the emission of one γ-ray:

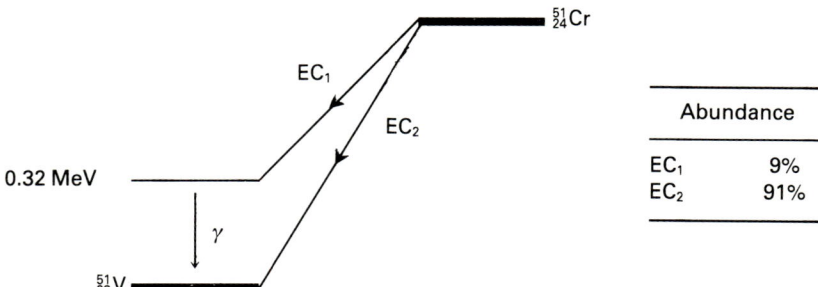

	Abundance
EC$_1$	9%
EC$_2$	91%

Positron decay does not occur with ^{51}Cr because the energy difference is only 0.75 MeV.

In some decay schemes the energy of the excited daughter is not emitted as gamma radiation, but is transferred to an inner orbital electron which is ejected. This process is called *internal conversion* and the emitted electron is called the internal conversion electron. The filling of the resultant vacancy gives rise to characteristic x-rays of the daughter or to Auger electrons. The probability of internal conversion taking place increases with increasing atomic number, the closeness of the energy of the transition to the electron's binding energy and the lifetime of the excited state.

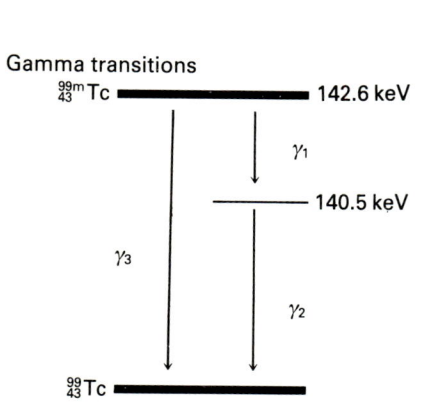

Gamma transitions

	Energy (keV)	Abundance
γ$_1$	2.1	0.99 (100% IC)
γ$_2$	140.5	0.99 (11% IC)
γ$_3$	142.6	0.01 (100% IC)
X-rays		
Kα	18	0.07
Kβ	21	0.01
IC electrons		
K	120	0.10
L	139	0.01
M	2	0.99
Auger	1	1.37

In a few decay schemes, the transition of the excited daughter to the ground state does not take place virtually instantaneously and the excited state has a lifetime ranging from seconds to hours. These long-lived excited states are called *metastable states*. When they decay, emission of gamma radiation and/or electrons due to internal conversion of the gamma radiation takes place. This type of decay is termed *isomeric transition*. An important radionuclide of this type is technetium-99m, the product of the ^{99}Mo–^{99m}Tc generator (§1.5.2). The gamma transitions in its decay scheme are shown at the foot of p 24.

2.2.5 Complex decay schemes

Many radionuclides have complex decay schemes. These may be due to the emitted particle having several possible maximum energies, leading to different energy levels in the daughter nucleus and to a number of gamma radiations of different energies. For example, iodine-132 has six different beta energies and 11 gamma energies with abundance of four per cent or more. Complex schemes may also be due to decay occurring by more than one of the above mechanisms. A combination of electron capture and positron emission is quite common, for example in cobalt-58 where 85% of transitions are by way of electron capture and 15% by positron emission. In some cases, e.g. copper-64 and rubidium-84, beta decay, positron decay and electron capture all take place and gamma radiation can accompany all three processes. Full descriptions of the decay schemes of the common radionuclides in nuclear medicine are given in Dillman (1969, 1970).

2.3 Interactions of Radiation with Tissue

2.3.1 Alpha particles

Alpha particles are emitted with high energy, typically between 3 and 7 MeV. Due to their heavy mass and double electronic charge, they interact strongly with surrounding atoms giving a very high ionisation density and linear energy transfer along their track. For example a 5 MeV α-particle creates 7000 ion pairs per mm at the start of its track (in tissue) and has a linear energy transfer of 55 keV μm^{-1}. This rapid loss of energy results in a very short range, typically less than 0.1 mm. The rapid deposition of energy also results in considerable biological damage. For both these reasons, radionuclides emitting α-particles are not suitable for use in nuclear medicine.

2.3.2 Electrons

Beta particles, positrons, Auger and internal conversion electrons have a range in tissue from a few μm to a few mm. However, the electron, in losing energy, follows a tortuous path and the mean range is typically a fifth of the maximum range. Energy losses are principally by excitation and ionisation, but the interaction is less strong than for α-particles because of the much smaller mass and single electronic charge. For example a 1 MeV β-particle creates about 20 ion pairs per mm at the start of its track and has a linear energy transfer of 0.25 keV mm^{-1}.

Beta emitting radionuclides, because of the limited range of the beta emission, are not generally suitable for use in *in vivo* studies with external detectors. Their use is confined therefore to studies where biological samples are taken, e.g. blood or urine, and the radioactivity can be assayed through close contact of the sample with the detection system, e.g. liquid scintillation counting (see §4.1). Positron emitting nuclides are suitable for studies with external detectors which respond to the pair of gamma rays emitted when the positron is annihilated. Auger and internal conversion electrons are absorbed close to their origin because of their very low energies. They cannot be detected easily, but of course contribute to the radiation dose.

2.3.3 Gamma and x-radiation

Gamma and x-ray photons are attenuated exponentially as they pass through matter and do not have a finite range like charged particle radiations, i.e.

$$I = I_0 e^{-\mu m}$$

where I_0 is the initial intensity of the radiation which is attenuated to an intensity I after traversing material with a mass per unit area m (g cm^{-2}) perpendicular to the radiation path and a mass attenuation coefficient μ (cm^2 g^{-1}). Attenuation is due to two causes; scatter and absorption. The relative effects of these two processes will vary with energy throughout the range 0.03–1.3 MeV met in nuclear medicine.

Scatter may be elastic (Thomson scattering) or inelastic (Compton scattering). In elastic scattering, the photon is scattered by an orbital electron without loss of energy. It is only important at low energies below those usually met in nuclear medicine. In Compton scattering, the photon energy is much greater than the binding energy of an orbital electron and part of its energy is transferred to the electron (the recoil electron) whilst the scattered photon has lower energy. After a number of Compton interactions the photon will have lost sufficient energy to enter the range where a photoelectric interaction will occur. The energy of the recoil electrons is absorbed locally by ionisation and excitation. High energy photons are scattered less than low energy ones and therefore the mass attenuation coefficient decreases with increasing energy. The attenuation coefficient is approximately constant with atomic number (Z). Compton scattering is the most important loss mechanism in the energy range 0.03–1.30 MeV.

Absorption can take place as a result of the photoelectric effect or pair production. In a photoelectric interaction, the photon interacts with an atom to overcome the binding energy of an electron and to eject it with kinetic energy. The kinetic energy of the photoelectron will be absorbed locally by ionisation and excitation. The vacancy in the electron shells of the atom will be filled by other electrons, with the emission of characteristic x-radiation. The mass attenuation coefficient due to the photoelectric effect decreases approximately as the inverse cube of energy. As a result it dominates the total attenuation at low energies and all materials show a rapid drop in absorption with increasing energy at low energies. The attenuation coefficient also increases with Z, so these effects are more marked in high Z materials, such as lead.

Beta–positron pair production takes place when a photon with energy greater than 1.02 MeV passes close to a nucleus. The energy in excess of 1.02 MeV is imparted to the two particles and is subsequently lost by ionisation and excitation. Eventually the positron is annihilated with the emission of two 510 keV gamma rays. Clearly absorption due to this cause only becomes important over 1.02 MeV and is not significant in the energy range met in nuclear medicine.

The total mass attenuation coefficient will be the sum of the attenuation coefficients corresponding to each of the above causes. The relative importance of the different mechanisms for low and high Z materials is shown in figure 2.1. A fuller treatment of these attenuating mechanisms is given by Johns and Cunningham (1974).

Much of the gamma radiation emitted during the decay of radionuclides in body organs will leave the body. As stated earlier, attenuation of the radiation is exponential and this is not too severe for the γ-ray energies commonly used in nuclear medicine. For the extreme range of energies the thickness of tissue which reduces the intensity by one half (the half-value layer) is 2 cm at 30 keV and 10 cm at 1.3 MeV. Above 100 keV, where the half-value layer is 4 cm, the attenuation introduced by body tissues is not serious in most measuring situations. The absorbed fraction of the gamma radiation clearly contributes to the radiation dose.

Ideally therefore, to minimise radiation dose, a radionuclide emitting high energy γ-rays is desirable since little of the radiation will be absorbed. Unfortunately the same process must apply to the detection system and low absorption will lead to a poor detection efficiency (see §3.2.1). As a result, a lower energy of gamma radiation is more generally optimal and represents a compromise between higher radiation dose and a more practicable detector efficiency. This situation is only achieved in practice with the radionuclides which decay by isomeric transition. The importance of technetium-99m is a consequence of its pure γ-ray emission with low radiation dose and at a suitable energy for efficient detection by sodium iodide scintillation detectors. However, as shown earlier, the emission of gamma radiation during the decay of many radionuclides is accompanied by the emission of charged particle radiation. The latter, whilst not contributing to the external detection properties of the radionuclide, often contributes substantially, because of its total absorption, to the radiation dose. It is often the factor which limits the dose of radioactivity that can be administered safely. The exact radiation dose will depend on the abundance of the particle spectrum.

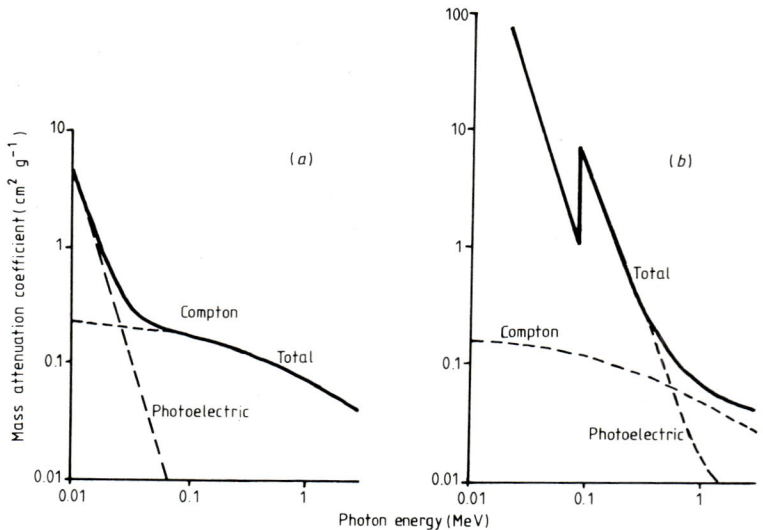

Figure 2.1   Variation of mass attentuation coefficient with photon energy for (a) water and (b) lead.

2.4 Radiation Damage

The value of the information to be gained from using a procedure requiring the administration of a radionuclide must always be balanced against the risk from the radiation dose. Energy will be absorbed by the body tissues from the radiation as a consequence of the processes described in the previous section. The unit of energy absorbed per unit mass of tissue is commonly the rad, defined as 100 erg g^{-1}. This unit is currently being replaced by the SI unit, the gray (Gy), defined as 1 J kg^{-1} (1 Gy = 100 rad).

The total radiation dose to an organ will be due to both the absorption of penetrating and non-penetrating radiation arising from within the organ and the absorption of penetrating radiation from outside the organ. The absorbed dose due to each source will depend on (a) the total number of disintegrations of the radionuclide in the source, (b) the energy emitted per disintegration, (c) the fraction of this energy absorbed by the target organ and (d) the mass of the target organ.

Using the terminology of the Medical Internal Radiation Dose Committee (MIRD) the mean absorbed dose D may be expressed (Brownell *et al* 1968) as:

$$D = \frac{A}{M} \sum_i \Delta_i \varphi_i \qquad (2.1)$$

where A is the integrated activity (e.g. Bq h) in the organ, M is the mass of the organ (g), Δ_i is the equilibrium absorbed dose constant and is the energy emitted per unit activity by component i of the radiation spectrum, and φ_i is the fraction of the energy emitted in component i which is absorbed by the target organ.

The first term (A/M) of equation (2.1) contains principally the biological determinants of the radiation dose. The integrated activity A will be a function of (a) the rate of physical decay and (b) the biological handling of the activity by the organ. An upper limit to the radiation dose can be calculated by assuming there is no biological clearance and the radioactivity in the organs is lost solely by radioactive decay. In general, however, biological clearance will play a part. If this is exponential, the overall rate constant, λ, is given by:

$$\lambda = \lambda_p + \lambda_b$$

where λ_p is the physical decay constant and λ_b is the biological rate constant.

The second term, $\Sigma_i\Delta_i\varphi_i$ of equation (2.1) contains the physical determinants of radiation dose. The equilibrium absorbed dose constant Δ_i is given by:

$$\Delta_i = 7.88 \times 10^4 n_i E_i \tag{2.2}$$

where n_i is the fractional abundance per disintegration of radiation with mean energy E_i (MeV). The constant 2.13 is the conversion constant from the basic units, MeV and disintegrations per second, to the derived units of rads and Bq h. φ_i is the absorbed fraction of the radiation component with mean energy E_i for particular target and source volumes at particular locations. Values of φ_i at different energies of penetrating radiation have been tabulated for a wide combination of body organs as sources and targets by Snyder *et al* (1969). In the case of non-penetrating radiation, φ_i equals unity when the target organ is also the source, but is zero for all other organs. This can be assumed to be true for practical purposes for all electrons and for photons of energy of 10 keV or less. The summation over $n_i E_i$ will clearly depend on the nature of the radiation spectrum of the radionuclide.

In evaluating the risk from radiation, the dose to the target organs, the whole body and other organs which may concentrate the radioactivity either during its metabolism or excretion must be considered.

The absorption of radiation by living cells always produces potentially harmful effects. These range from death after a few days for large doses to the induction of cancer 20 years later for a low dose. The damage originates from chemical changes in the complex organic molecules in cells. These changes take place rapidly in a few seconds and are the consequence of reactions with the very reactive H and OH free radicals formed by the radiolysis of water, the most abundant cellular constituent, i.e.

$$H_2O \rightarrow H_2O^+ + e^-$$

$$H_2O + e^- \rightarrow H_2O^-$$

$$H_2O^+ \rightarrow H^+ + OH$$

$$H_2O^- \rightarrow OH^- + H.$$

The chemical changes substantially alter the biological properties of the organic molecules, e.g. RNA, and over a longer period of time these may alter the structural and functional properties of the cells. This can result in premature cell death, the prevention or delay in cell multiplication and the introduction of inheritable defects in cells. Biological defects induced by radiation are divided into two groups, somatic and hereditary effects. Somatic effects are those which affect only the organism irradiated. Hereditary effects affect future generations only and arise from damage to germ cells in the reproductive organs.

It has been found from experience that the severity of radiation damage is related not only to the radiation dose, but also the dose rate and the type of radiation. This is expressed for the purpose of risk in the concept of dose equivalent, E, which is given by the expression:

$$E = DQN \qquad (2.3)$$

where D is the mean absorbed dose, Q is the quality factor of the radiation and N is a modifying factor based on dose rate. The commonly used unit of equivalent dose is the rem ($100 \, \text{erg} \, \text{g}^{-1}$) where D is in rads, but is being replaced by the SI unit the sievert ($\text{J} \, \text{kg}^{-1}$) where D is expressed in grays. At present N is given the value of unity. Q takes into account the distribution of absorbed energy and is related to the linear energy transfer (§2.2). When the radiation spectrum and hence the energy transfer cannot be specified everywhere, Q takes the value of unity for electrons and x and gamma radiation, 10 for neutrons and protons and 20 for alpha particles. A more detailed discussion of health risks due to irradiation is given by the International Commission on Radiological Protection (1977).

In practice the radiation doses to critical organs and/or the whole body from doses administered for diagnostic purposes are well below the levels at which obvious harmful effects are observed. In the United Kingdom, doses of radioactivity are limited so that the whole body dose does not in general exceed 5 mSv (0.5 rem) or the critical organ dose exceed 25 mSv (2.5 rem) (Notes for Guidance 1979). This compares with an annual radiation dose to the whole body of 1 mSv received from natural sources and the maximum permissible whole body dose of 50 mSv for occupationally exposed persons (Code of Practice 1972).

3 Solid Scintillation Counting

3.1 Introduction

The most important clinical method of detecting radionuclides is scintillation counting. This relies on the ability of ionising radiation to cause fluorescence (scintillations) in phosphors by excitation. The intensity of light in each scintillation is proportional to the total energy absorbed by the phosphor from each γ-ray or β-particle. The scintillations are converted by one or more photomultiplier tubes to electrical pulses which can be processed and counted.

Solid scintillation counting is used to measure the radioactivity of radionuclides which decay with the emission of γ-rays or very energetic β-particles. A solid scintillator is required to absorb the radiation and provide efficient detection. This technique is often referred to as 'gamma counting'.

3.2 The Solid Scintillation Detector

3.2.1 Principle of operation

A typical solid scintillation detector is shown in figure 3.1. A γ-ray entering the crystal ionises an atom or number of atoms and the secondary electrons produced excite atoms along their paths. The excited atoms return promptly to their ground state with the emission of ultra-violet light. The most widely used solid scintillator is a crystal of sodium iodide, usually containing a small amount (0.1–0.4%) of thallium, which enables scintillation to take place at room temperature. This is often abbreviated to NaI(Tl). Sodium iodide has most of the properties required for a good radiation detector. These are:

(a) It is a good absorber of gamma radiation due to its high density (3.67 g cm^{-3}) and high atomic number.

(b) The scintillation intensity is proportional to the energy absorbed over a wide range of energies and enables the energy to be measured.

(c) It is transparent so that the scintillations can reach the photo-cathode without attenuation.

(d) The scintillations are of short duration so that the dead time of the detector is low. In practice, 90% of the light output occurs in about 800 nanoseconds, giving a dead time of about one microsecond.

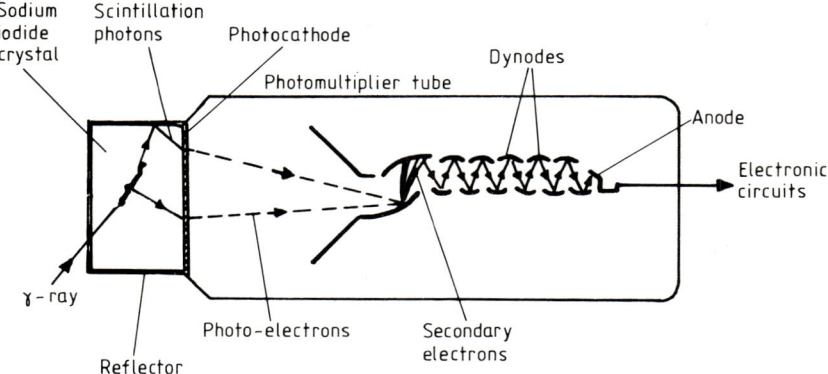

Figure 3.1 Solid scintillation detector.

One face of the crystal is optically coupled to the photocathode of the photomultiplier tube. This is usually done with silicone grease. The photocathode is comprised of rare earth compounds and has a low quantum efficiency, between 5 and 10 photons being necessary to release one photoelectron. These photoelectrons are accelerated in the high vacuum of the tube towards the first dynode by a positive voltage applied to it. The dynode is an electrode coated in a material similar to the photocathode. On impact each accelerated electron will stimulate the emission of two or three electrons. These new electrons are then accelerated towards the second dynode by a higher voltage applied to it and the amplification process occurs again. The photomultiplier will typically have 10–12 dynodes and amplification will occur at each giving an overall gain of 10^6–10^7. The arrival of the bunch of electrons at the anode gives rise to a current pulse, which can be processed by external circuitry. The anode–cathode voltage is usually in the range 1000–1500 V and is distributed evenly along the dynode chain by a resistive network.

Sodium iodide is hygroscopic and turns yellow on exposure to water vapour. The crystal is therefore hermetically sealed in a thin-walled aluminium case except where it joins the photocathode. The case has a diffuse reflecting inner surface to direct the scintillations towards the photocathode and also serves to prevent the entry of light to the photocathode.

3.2.2 Pulse height spectra

Not all the energy of the incident γ-ray will necessarily be absorbed in the crystal. This will depend on the type of interaction which takes place to produce the secondary electron or electrons, namely photoelectric, Compton or pair production. At low energies, the photoelectric effect will dominate and the incident γ-ray will transfer all its energy to the secondary electron, whose final energy will be that of the γ-ray minus the electron's binding energy (which will usually be the 29 keV binding energy of the K shell of iodine which is the most likely source of the photoelectron). The energy of the photoelectron will be absorbed in the crystal and in most cases the associated characteristic iodine K x-rays will also be absorbed. This will give rise to an electronic pulse with a single amplitude corresponding to the total energy of the γ-ray. In small crystals, the characteristic x-rays may escape from the crystal and the amplitude of the electronic pulse will be a little lower than that corresponding to the total energy. This is only noticeable at low γ-ray energies, below 100 keV.

At higher energies, Compton interactions become important. The energy of the Compton electron is absorbed and the Compton γ-ray may escape from the crystal or take part in further interactions. If the γ-ray leaves the crystal, a range of pulse heights is observed corresponding to the range of energies which can be taken by the Compton electron up to a maximum (the Compton edge). This is likely in small crystals. Alternatively the γ-ray may suffer a series of Compton interactions. The total energy transferred to the secondary electrons is additive and pulse heights corresponding to energies greater than the Compton edge and almost reaching the pulse height corresponding to the total energy are observed. If a series of Compton interactions end in a photoelectric interaction, the total energy of the incident γ-ray is absorbed.

At γ-ray energies above 1.02 MeV, pair production becomes possible with absorption of the energy of the electron and the production of two 0.51 MeV γ-rays from positron annihilation. The two annihilation γ-rays may escape from the crystal, one may escape or both may be absorbed. The first situation gives rise to the 'double escape peak' 1.02 MeV less than the incident energy, the second to the 'single escape peak' 0.51 MeV less than the incident energy and the third to a single peak corresponding to total absorption. Compton interactions will provide a continuum of pulse heights between these peaks. Typical spectra for low energy (151 keV) and medium energy (662 keV) γ-rays are shown in figure 3.2.

A number of secondary effects also contribute features to the shape of the spectrum. Gamma-rays interact with the source material producing Compton scattered γ-rays (back-scatter). For detection, the scattering angle has to be large. This results in a narrow range of energies and since absorption is by the photoelectric effect a well defined peak can occur. Gamma-rays will also interact with the lead shielding around scintillation detectors with the emission of characteristic x-rays. The 73 keV characteristic x-radiation from the K shell of lead may be observed. Complex spectra will be observed when a radionuclide emits more than one energy of gamma-radiation.

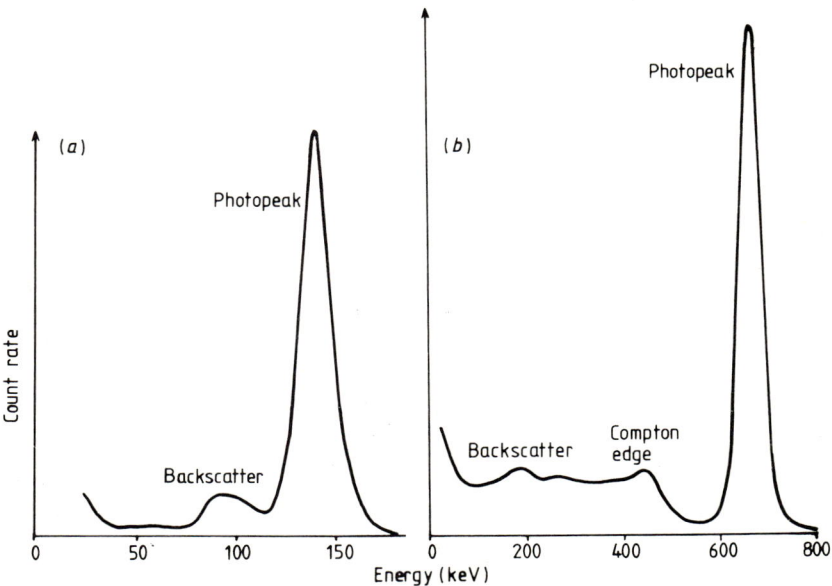

Figure 3.2 Measured γ-ray spectra for (a) ^{99m}Tc and (b) ^{137}Cs.

It will be seen from figure 3.2 that there is no single pulse height corresponding to the total energy of the incident γ-ray but a range of pulse heights. This is called the 'photopeak' and its finite width arises because of statistical variations which occur in the processes converting the energy of the incident γ-ray to an electronic pulse. These variations occur throughout the pulse height spectrum and arise as follows:

(a) The production of light in the crystal varies with the ionisation density of the secondary electron track and the number of secondary electrons produced.

(b) The light reaching the photocathode depends on the position of the interaction within the crystal.

(c) Production of photoelectrons at the photocathode and at each of the dynodes is a statistically variable process.

(d) Electronic instabilities will give rise to variations in the pulse height.

A measure of the broadening of the photopeak by these processes is given by the energy resolution. This is defined as the full width of the photopeak at half maximum (FWHM) and is expressed as a percentage of the pulse height of the maximum. Under good conditions, using a single crystal and selected photomultiplier tube, a resolution of 7% can be achieved at 662 keV.

Fuller treatments of solid scintillation counting are given by Birks (1964) and Hine (1967).

3.2.3   Counting conditions

In practice it is unusual to count all the pulse heights present in a spectrum such as those shown in figure 3.2. Sections of the spectrum may be selected for counting (see §3.3.4) and it is normal to count only pulses occurring in the photopeak. This results in a lowering of the measured count rate, but increases the sensitivity over the background count rate. The background spectrum generally drops with increasing pulse height but can have an appreciable count rate over the entire spectrum corresponding to the radionuclide. By counting only the photopeak region, the count rate in a limited range of pulse heights is maximised for the sample and minimised for the background. Counting pulses in the photopeak only also excludes lower energy pulses arising from back-scatter and characteristic x-rays from lead shielding. This is most important in devices which image a radionuclide distribution because the count rate then corresponds to the true radioactivity at each point in the image and the scattered radiation which is due to radioactivity elsewhere is excluded.

3.3   Electronic Modules

3.3.1   General arrangements

The electrical pulses from a solid scintillation detector generally need to be amplified and undergo pulse height analysis to remove extraneous pulses before they can be processed. The processing will differ according to the type of measurement being made. A scaler/timer is employed when constant radioactivity (excepting the physical decay) is being measured. This may be a radioactive sample in a container or radioactivity in an organ or region of the body which is appreciably constant over the measurement period. A ratemeter is used to record the variation in radioactivity when rapid changes are taking place. The electronic modules required to make up each of these counting systems are shown in figure 3.3. The function of each module is described in outline below.

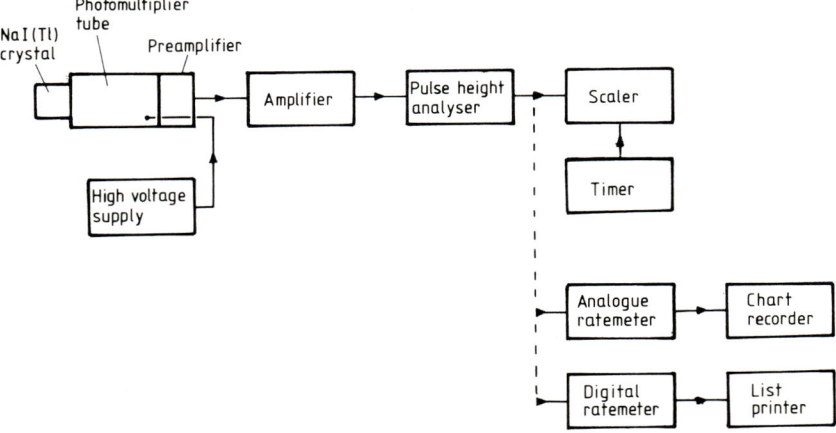

Figure 3.3   Electronic modules for counting systems.

3.3.2   Pre-amplifier

The scintillation detector and the electronic equipment are normally connected by a flexible cable so that the detector can be moved around. To overcome the attenuation introduced by the cable, a pre-amplifier is often connected to the output of the detector and is an integral part of the detector assembly. An emitter follower pre-amplifier is usually employed to match the high impedance (approximately 1 MΩ) of the photomultiplier tube to the low impedance of the cable. The amplitude of the pulses in the cable is typically a few mV.

This arrangement requires the provision of a low voltage supply to the detector assembly in addition to the high voltage supply and the signal line. A simpler arrangement can be employed if a charge sensitive amplifier (§3.3.3) is employed. A single coaxial cable is used to carry the high voltage supply to the detector and the pulses to the amplifier. The pulses are decoupled from the high voltage at the amplifier input by a capacitive connection.

3.3.3 Amplifier

Voltage sensitive or charge sensitive amplifiers may be used. The amplifier has a gain of up to 1000 to increase the amplitude of the pulses to between one and 10 volts. To maintain the relationship between pulse height and radiation energy, the amplifier must have a linear response and the gain must be stable against fluctuations in line voltage and temperature. It must also be insensitive to overloads from large pulses and high count rates. The former arise from background radiation of greater energy than that of the nuclide being counted. The latter give rise to 'pulse pile-up' when pulses become superimposed on the trailing edges of previous pulses and distortion of the pulse height spectrum occurs. The amplifier shapes the pulses to a standard form. It has a rise time of about 0.2 μs in order to respond to all pulses from the detector. The decay time is trimmed so that pulse duration (from start to 10% of peak amplitude) is about 1 μs. Under these conditions, a count rate of 10^5–10^6 counts per second can be handled without losses.

3.3.4 Pulse height analyser

The pulse height analyser discriminates between pulses in the range of measurement and extraneous pulses. It has two modes of operation, integral mode (or pulse height discrimination) and differential mode (or pulse height analysis). The second is more commonly used.

In integral mode, all pulses with amplitudes greater than a set height are accepted for further processing. This facility is useful for rejecting low energy pulses arising from electronic noise.

In differential mode, only pulses with amplitudes within a set range of amplitudes (often called the 'window') are accepted. This is useful for reducing the count rate due to background and scattered radiation to that portion which happens to have pulse heights within the window.

On counting instruments, the minimum pulse height for integral mode operation is set by a continuously variable 'discriminator' or 'threshold' control. Selection of the 'window' for differential analysis varies from instrument to instrument. One type employs independent variable minimum and maximum height controls, a second has variable threshold and window width controls and a third has variable window centre-line and width controls with the window width graduated as a percentage of the centre-line pulse height. These arrangements are illustrated in figure 3.4.

For ease of calibration and use, the threshold or centre-line controls should be linearly related to energy. Linearity is sometimes defined as the maximum deviation from such a linear relationship expressed as a percentage of the setting at which it occurs. Stability of the analyser's threshold or centre-line and window controls is also important for good performance. Each pulse accepted by the analyser triggers the output of a monoheight pulse (up to 10 V in amplitude) with a very fast rise time, less than 100 ns, for further processing.

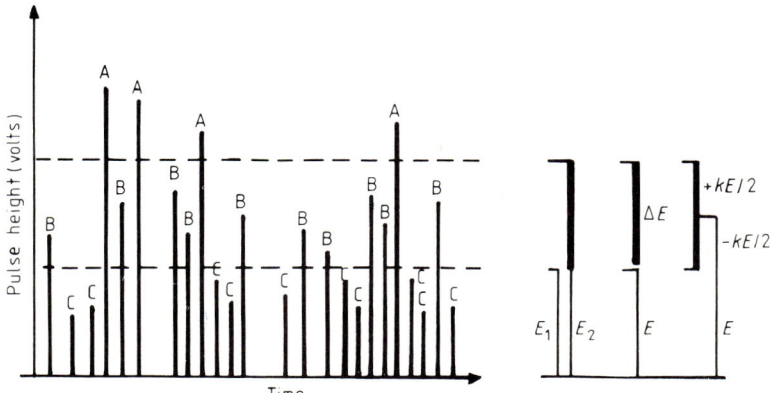

Figure 3.4   Pulse height analyser window arrangements. The pulses (B) in the window are accepted for further processing whilst those above the window (A) and below the window (C) are rejected. The window width may be selected by independent upper and lower controls E_1 and E_2, by a lower control E and a window ΔE, or by a window centre setting E and fractional width k.

The exact relationship between pulse height setting on the analyser and the energy of incident γ-rays will alter with the gain of the photomultiplier tube and the amplifier. The high voltage of the former and the gain of the latter are adjusted to give a suitable setting. Often, for operating convenience, the gain is adjusted so that the threshold or centre-line controls read directly in keV. More than one analyser can be connected to an amplifier to provide multiple windows on the pulse height spectrum.

3.3.5   Scaler/timer

A scaler/timer is used when the count rate is constant over the period of measurement. Its name derives from the earliest devices which used a mechanical register to record the number of pulses and required a cascade of 'scale of two' circuits to reduce the actual count rate to a rate which could be handled by the register. The pulses from the pulse height analyser are counted for a selected period of time or until a selected number of counts is reached, when the elapsed time is noted. The timer is essentially another scaler which counts the pulses from a mains frequency or crystal controlled oscillator. The selection of the preset count and time will depend on the accuracy with which the measurement is required (§3.8). The count and elapsed time can be read from a decimal display, both during and at the completion of counting. When several pulse height analysers are employed, a scaler will be necessary for each one to record the number of pulses in each section of the spectrum covered.

3.3.6   Ratemeters

Ratemeters provide a continuous indication of the pulse rate from a detector, and are used to record changes in radioactivity usually over a period not greater than 30 minutes. Two types of ratemeter are used; the analogue ratemeter and the digital ratemeter.

The analogue ratemeter is simpler and more common. Pulses from the pulse height analyser are impressed upon a capacitor C which discharges through a shunt resistor R. At a constant pulse rate, the charge on the capacitor is proportional to the pulse rate. The time to reach this equilibrium state is governed by the time constant RC of this circuit and this limits the response of the ratemeter to rapid changes in pulse rate. However the time constant also smoothes out the statistical noise (§3.8) in the count rate, with greater smoothing as the time constant increases. The value of the time constant has to be selected therefore as a compromise between smoothing of the noise and distortion of the count rate variation. The output of the capacitor–resistor circuit is displayed on a voltmeter calibrated in count rate as an instantaneous indication of count rate. A suitable impedance output is also provided to drive a potentiometric chart recorder to give a continuous record of the count rate variation with time. As with all analogue devices, the zero position and linear response need to be checked regularly for drift.

The digital ratemeter can be considered as a recycling scaler/timer with virtually no delay between one count and the start of the next.

This can be achieved in two ways. The usual way uses a buffer store. Pulses are counted in the scaler for a preset time and then the count is transferred to a buffer store. The scaler and timer are reset to zero and counting starts again. The time to transfer the count and to reset the scaler/timer is typically less than 10 μs. The count can be read from the buffer store relatively slowly whilst the next count is in progress. The alternative way employs two scalers and a timer. The pulse input is connected alternately to the two scalers, each of which is read whilst the other acquires counts. This is sometimes called the 'double buffer' method.

The total count being acquired by the scaler in the single buffer ratemeter or the active scaler in the double buffer ratemeter is usually given visually on a decimal display. The successive counts from the scalers may be recorded on a list printer or punched onto paper tape for subsequent processing. Digital ratemeters have two advantages over their analogue counterparts, namely their freedom from drift and their immediate response to varying pulse rates.

3.3.7 Power supplies

Low voltage and high voltage supplies are required for all counting systems. A low voltage stabilised DC supply (usually ±24 V) is necessary to power the pre-amplifier, amplifier, pulse height analyser and scaler or ratemeter.

The high voltage supply provides a DC voltage for the photomultiplier tube, usually adjustable in the range 1000–1500 V. It must be extremely stable with a variation less than 0.01% since a 1% change in voltage leads typically to a 10% change in gain and therefore pulse height. For the same reason, the AC ripple on the high voltage should be less than 1 mV peak to peak.

3.3.8 Multichannel analyser

A pulse height spectrum for a radionuclide can be obtained using a pulse height analyser and a scaler by making a series of measurements using a narrow window with increasing threshold values. This is obviously a tedious and repetitive process especially if the resulting spectrum indicates that the counting conditions, for example the gain, have to be changed and the spectrum rechecked.

A more convenient and rapid method is to use a multichannel analyser, which performs the counting in the successive channels in parallel over the same period of time. The multichannel analyser at its most basic comprises an amplifier, an analogue to digital converter, a data store and some form of data output. The data store typically has 256, 512 or 1024 memory locations (usually called channels) which cover the energy range as a series of contiguous narrow windows. The energy range will be set by the amplifier and conversion gains, together with a low energy cut-off to remove electronic noise. Each pulse height is converted to a digital value by the analogue to digital convertor and this becomes the address of one of the memory locations. Unity is then added to the contents of the addressed location indicating the collection of another pulse whose amplitude falls into the narrow energy range corresponding to that location. As counting proceeds, a histogram showing the frequency of occurrence of each pulse height is obtained. This spectrum of pulse heights can usually be transferred channel by channel to a printer or punched paper tape for further processing.

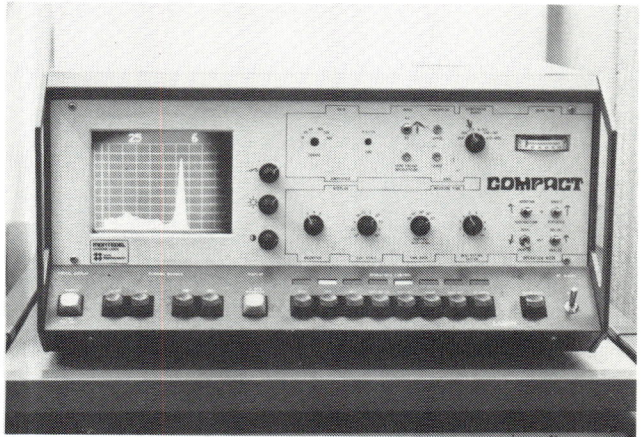

Figure 3.5 Compact multichannel analyser showing spectrum display (left), counting controls (upper right) and operating controls (lower right). (Reproduced by kind permission of EMI Nuclear Enterprises Ltd.)

Most multichannel analysers also have the facility to integrate the counts in a series of channels over a specified range, for example the photopeak. This enables them to be used as multiple scalers covering a number of energy windows, which can be selected accurately from the measured spectrum. A modern compact analyser is shown in figure 3.5.

3.4 Sample Counters

Different counting applications require different configurations of the detector and electronic modules described in § 3.2 and § 3.3.

3.4.1 Well counters

Sample counters measure a fixed amount of radioactivity (excepting its physical decay) in a container, e.g. from sequential blood samples. The simplest arrangements comprise a scintillation detector and a scaler. The detector has 'well geometry' in which the sodium iodide crystal has a blind hole to take the sample. In this way, the sample is encompassed as much as possible by the detector and its radioactivity counted as efficiently as possible. The detector is surrounded almost completely by a lead shield to reduce the entry of background radiation. This arrangement is illustrated in figure 3.6(*a*). The detector is usually cylindrical with a diameter of 50 mm or 75 mm and a height of 50 mm or 75 mm respectively. The well size should be only a little larger than the sample container to keep the crystal thickness as large as possible.

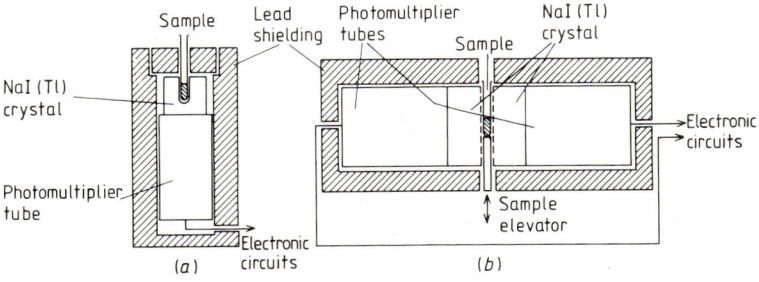

Figure 3.6   Sample counting arrangements. Elevation sections through (*a*) a well counter and (*b*) a radial hole crystal.

The amplifier, pulse height analyser, scaler/timer and power supplies are usually contained in a single case. A commercial example of this type of 'well' counter is illustrated in figure 3.7. Samples are placed manually in the detector one at a time. They are counted according to the preset conditions and the counts and/or elapsed time recorded manually from the visual display.

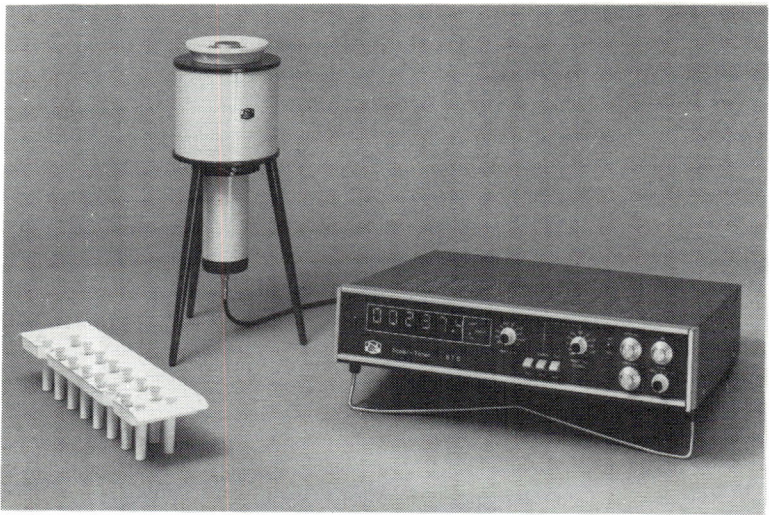

Figure 3.7 Well counter and scaler. A rack holding sample tubes is shown left, the well counter centre and the scaler/timer right. (Reproduced by kind permission of EMI Nuclear Enterprises Ltd.)

3.4.2 Automatic sample counters

The well counter is only suitable for small batches of samples since it requires someone to operate it all the time it is being used. For larger numbers of samples, e.g. from a radioimmuno-assay, automatic gamma counters which can hold between 100 and 600 samples are used. These comprise an automatic sample changer, a detector, one to three scalers, a timer and a data output device. The sample changer may have an endless belt or a system of moving trays to hold the samples. The latter has the advantage that the trays can be loaded in the laboratory away from the counter with a reduction in the counter loading time. In both systems when a sample reaches the counting position, it is mechanically loaded into the detector and is counted according to the preset conditions on the scalers and timer. At the end of the counting period, the sample number, counts and elapsed time are transferred to a printer or punched paper tape, the sample returns to the storage system and its place in the detector is taken by the next sample. The process takes place automatically until the batch of samples is finished.

Detectors have either the 'well' configuration described in § 3.4.1 or the radial hole arrangement shown in figure 3.6(*b*). In this latter arrangement a cylindrical sodium iodide crystal is placed between opposing photomultiplier tubes and has a radial hole which takes the sample container. For the well detector, the sample is either lowered to the bottom of the well by a grab mechanism or the detector assembly is raised pneumatically around the sample at the counting position. In the radial hole detector, the sample is lowered on an elevator. This arrangement has the advantage that the stopping position of the elevator can be altered to take account of different sample volumes and to optimise counting efficiency. The stopping position is normally set so that the sample is centred within the height of the crystal. Good detector geometry is also achieved by summing the outputs of the two photomultiplier tubes. In general, the radial hole counter copes better with counting larger volume samples than the well counter but relies on reproducible positioning of all the samples by the elevator. This is not necessary with a well counter since the samples rest on the bottom of the well. The detector assembly requires lead shielding to reduce the entry of background radiation. This is particularly critical between the detector and the sample storage area, so that the movement of active samples does not introduce a change in the background count rate from sample to sample. An example of a commercial automatic gamma counter is shown in figure 3.8.

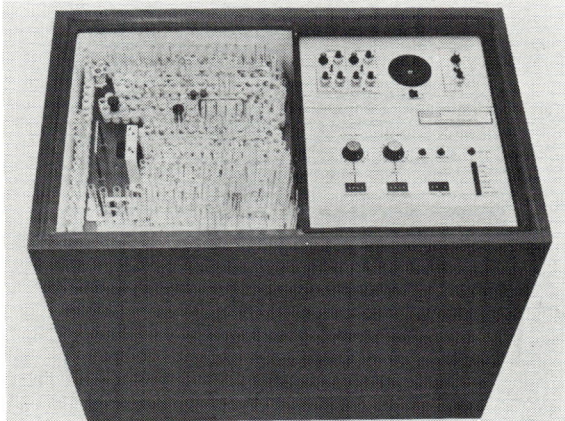

Figure 3.8 Automatic gamma counter. The continuous belt sample changer can be seen on the left with the counting position centre left. The spectrometer controls are top right and the counting controls bottom right. (Reproduced by kind permission of Packard Instruments Ltd.)

3.4.3 Large volume counters

The sample counters described in §3.4.1 and §3.4.2 take samples of small volume, typically 5 cm^3 or less and at most 20 cm^3. These volumes are acceptable if the concentration of radioactivity is high. However if the concentration is low it is often necessary to count the whole sample, e.g. a urine or faecal collection, to obtain an accurate result. Large volume counters measuring samples with volumes of up to one litre are necessary for this purpose.

It is important in counting a large sample to achieve good counting geometry so that the measured count rate is independent of sample position when the sample contains a non-uniform distribution of radioactivity. This is often the case with faecal collections which are difficult and unpleasant to homogenise. A number of arrangements to achieve good counting geometry have been described and three common ways are illustrated in figure 3.9. These employ firstly a container whose shape distributes the sample around a central detector, secondly a ring of detectors whose outputs are summed and thirdly a turntable arrangement on which the sample makes one revolution past a fixed detector. The last arrangement is most convenient using a standard shape of container, the least expensive with one detector and is available commercially.

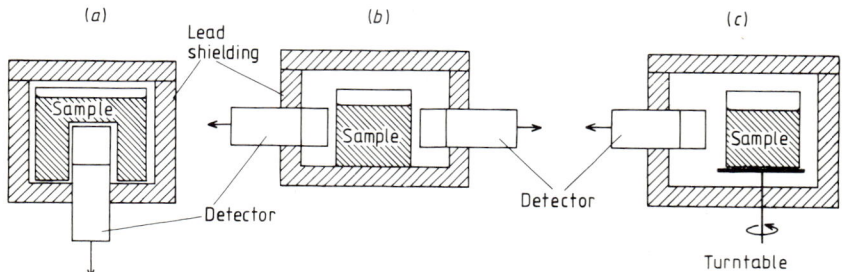

Figure 3.9 Large volume sample counting arrangements. The sample container may (*a*) surround the detector, (*b*) be surrounded by a ring of detectors or (*c*) rotate past a single fixed detector.

3.5 *In vivo* Uptake Counters

Uptake counters are used to measure the radioactivity in a region of the body, usually a particular organ. The radioactivity may be appreciably constant over the period of measurement or varying as a result of blood flow or a metabolic process. The counter comprises a solid scintillation

detector in a collimated shield and a scaler or ratemeter. The collimator
is made of lead and limits the direction from which radiation can reach
the detector to a particular field of view. Radiation from other directions
is absorbed. The detector and shield are usually mounted on a mobile
stand for flexibility in positioning over the patient.

The scintillation detector usually employed for *in vivo* counting has
a 50 × 50 mm cylindrical NaI(Tl) crystal. The two types of detector
collimation are commonplace; parallel hole and diverging hole collimators. These are illustrated in figure 3.10.

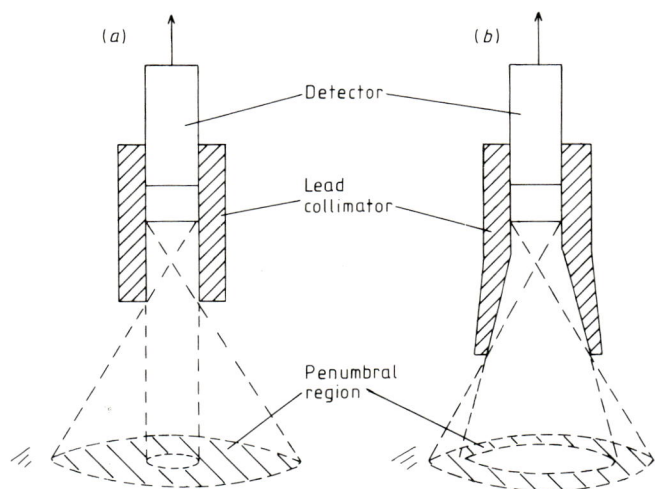

Figure 3.10 Collimators for *in-vivo* counters. (*a*) Parallel hole collimator and (*b*) diverging collimator (right).

Parallel hole collimation is employed when it is desirable to retain a
constant field of view with depth in tissue. This is necessary when
measuring the radioactivity in deep organs such as kidney or spleen. A
penumbral zone exists around the geometrical field of view and this
enlarges with distance from the collimator. The penumbra can be
minimised by increasing the depth of collimation and this is important
for reducing the contribution from neighbouring tissue. However this
also has the effect of increasing the detector to organ distance with a
loss of sensitivity. The length of collimation is therefore a compromise
between increased sensitivity and reduced penumbra. For maximum
sensitivity, the detector should always be placed as close to the skin as
possible.

A diverging hole collimator is used for regions which subtend a large solid angle at the detector when the detector is placed at a reasonable distance from the organ for good sensitivity. The geometrical field of view increases with distance from the detector and a constant detector–skin distance must be employed for reproducible sensitivity. The diverging collimator has the advantage that for a given geometrical field of view the penumbra is smaller than that of the parallel hole collimator. There is also less penetration of the collimator by radiation originating from just outside the field of view due to the greater path length in the lead collimator. This also helps to reduce the penumbra. The commonest application of the diverging hole collimator is measurement of the thyroid uptake of radionuclides (§ 6.4.3).

3.6 Whole Body Counters

As their name implies, whole body counters measure the total radioactivity in the body. In clinical studies, they are used principally to measure variation in the retention of an administered radionuclide with time. This enables a direct and quick measurement instead of the laborious collection and counting of excreta, whose total activity has to be subtracted from the dose given to yield the radioactivity retained. The radioactivity in the patient may be calibrated by reference to a body phantom containing a known activity or, more commonly, expressed as a percentage of the dose, measured by counting the patient soon after its administration before significant excretion has occurred.

The principal requirements for a whole body counter are (a) sensitivity, in some instances to body contents around 40 Bq (1 nCi), and (b) a counting response which is independent of the distribution of the radionuclide in the body. There are a number of ways of achieving these objectives and the commonest are the 'shielded room' and the 'shadow shield' counters.

3.6.1 Shielded room whole body counters

In this arrangement, the counter and patient are inside a room having massive shielding against background radiation to aid sensitivity. The room is usually made of steel 100–200 mm thick and is sometimes surrounded by concrete and earthworks. The latter provides satisfactory shielding for high energy photons and the inner steel serves as a barrier against low energy scattered photons. The contribution of the shielding itself to the background count rate is minimised by using pre-1939 steel and low activity minerals.

Large scintillation detectors are employed in the majority of whole body counters and these are deployed in a number of ways to achieve sensitive and uniform counting of the radioactivity in the patient. The commonest way employs a large number of fixed detectors in two layers, one above and the other below the patient. For example, the counter at the Brookhaven National Laboratory (Cohn *et al* 1969) employs two sets of 27 detectors arranged in a 3×9 matrix, each detector being 150 mm in diameter and 50 mm thick. A second arrangement is to use a ring of detectors surrounding the patient, which can travel the length of the patient. Alternatively the patient can travel on a moving couch completely through the ring but this doubles the required length of the shielded room and greatly adds to the cost.

This type of body counter can be readily used to measure natural ^{40}K in the body which has a typical activity of 7.5 kBq (0.2 µCi). It is also often used in a radiological protection role to monitor the results of the accidental ingestion of radioactive material. Care must be taken not to accidentally contaminate shielded rooms.

3.6.2 Shadow shield whole body counters

For most clinical investigations, the high sensitivity of the shielded room counter described above is unnecessary and simpler detector arrays and less shielding (with consequent higher background count rate) are adequate. This results in counters which are lighter and less expensive and therefore easier to include in hospital designs.

The design of many hospital whole body counters relies on the 'shadow shield' principle and the counter is sited in an ordinary room. The principle is illustrated schematically in figure 3.11. One or two detectors are sited above and below the patient who lies on a horizontal couch which moves between them for his entire length. This typically takes 15 min. Scintillation detectors with diameters ranging from 100 to 150 mm and thicknesses from 75 to 100 mm are used. The detectors are contained in a lead shield, typically with 100 mm thick walls, and their field of view is collimated to a strip across the patient about 250–300 mm wide. Additional lead sheet 50 mm thick extends beyond the central shield to cast a 'radiation shadow' over the opposing detector, so that no radiation can directly enter each detector. Scattered radiation will be able to reach the detectors. This increases when a scattering medium, for example the patient, is present. However since the scattering angle must be 90° or more for simply scattered photons, the scattered γ-rays will have a maximum energy of 0.51 MeV. As a result the count rises preferentially in the lower energy range (0.1–0.5 MeV) and is generally higher than in comparable shielded rooms.

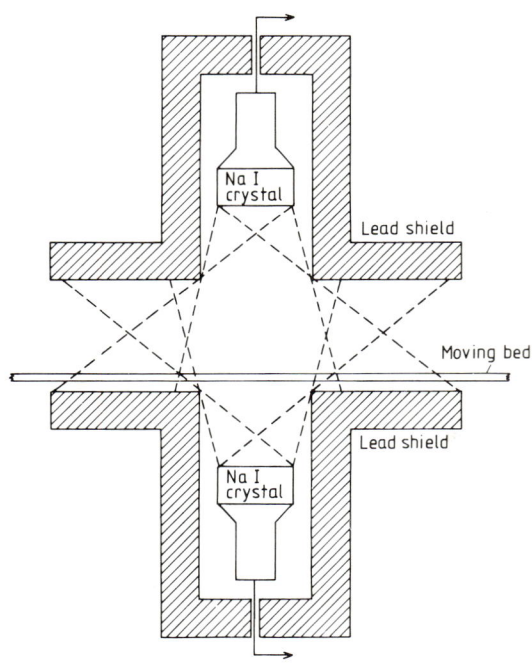

Figure 3.11   Shadow shield arrangement for whole body counter. The extremes of the field of view of each detector are covered by the opposite shield.

The total weight of shielding in a counter of this type is about 5 tonnes. By comparison a small shielded room with dimensions $2 \times 2 \times 2$ m and 75 mm thick lead walls would weigh about 22 tonnes.

Shadow shield counters can measure body activities down to about 3.5 kBq (0.1 μCi) and can have a ±2% variation in response when wide spectrometer windows accepting both photopeak and scattered events are used. Generally a variation in response of ±5% is acceptable for clinical investigations. Shadow shield counters are more versatile than shielded rooms. Most can be fitted with detector collimation to reduce their sensitivity and enable profile scans along the length of the patient to be performed. Their use can then also be extended to studies with MBq (mCi) doses of radioactivity, e.g. [131]I, to locate metastatic deposits from thyroid cancer. Shielded rooms cannot usually be adapted in this way and saturation of the count rate occurs with high activities.

A listing of whole body counters, with details of detector layout, and performance characteristics is given in the *Directory of Whole Body Counters* (1970) and more recent developments by Sorenson (1974).

3.7 Imaging Devices

Radionuclide imaging systems measure the distribution of radioactivity in an organ or region of the body. Such systems are the rectilinear scanner and the gamma camera and are primarily used for imaging static distributions in brain, liver, skeleton, lung, heart, etc, for diagnostic purposes. As indicated in the Preface, these applications, although important and commonplace, are not the subject of this monograph. However, imaging devices are also increasingly being used to investigate metabolism and blood flow through the quantification of serial images showing the movement of radioactivity. Brief descriptions of the scanner and gamma camera and the capture and processing of the serial data are therefore appropriate.

3.7.1 Rectilinear scanner
During scanning the patient lies supine beneath a solid scintillation detector with a focussed multihole collimator. This arrangement responds principally to gamma radiation originating from a small volume of tissue (typically 10^3 mm^3 in volume) between 100 and 150 mm directly beneath it. The detector is driven in a raster pattern in a horizontal plane over the area of the patient to be imaged. A ratemeter system (figure 3.3) measures the variation in count rate observed as a result of the movement over the radionuclide distribution. The output of the ratemeter is recorded by an output device moving in synchrony with the detector and this maps out the distribution of count rate and hence radioactivity. Various output devices have been employed. The earliest used a light source to darken x-ray film. The intensity of the light source increased with count rate and the darker areas on the film therefore correspond to the more radioactive areas. Latterly mechanical dot-printers have been used to mark a sheet of paper. In the simpler systems, a mark is made on the paper for each count or set number of counts and the density of marks reflects the distribution of radioactivity. This format

is limited in the range of count rate information that can be provided before the marks become indistinguishable. In more complex systems, the ratemeter signal is also used to drive a carriage carrying a multicolour ribbon (often with eight colours) beneath the printer. The colours of the rainbow are then used to indicate different levels of count rate, the red end conventionally being used to represent the most active areas of the distribution. This format conveys a wider range of count rate information than the monochrome printer.

Many later models of scanners had two detectors on a common vertical axis; one above the patient pointing down and the other beneath the patient pointing up. The outputs of the two detectors could be summed to provide a sensitivity response which is almost independent of depth in tissue.

To quantify the radioactivity in some part of the distribution, it is necessary to count the dots in the area of interest. This is either done manually or using a scaler. For manual counting, the factor setting the number of counts per dot must be selected so that the dots are clearly distinguishable even at the highest count rates. The variation of radioactivity with time can be determined by counting the dots in an identical area in a series of scans. These activities can be expressed in terms of the administered dose by scanning a known aliquot of the dose under the same geometrical conditions as the anatomical feature of interest. The count due to a dose can then be calculated from the number of dots in the area covering the aliquot times the factors relating counts to dots and the dose to the aliquot.

This technique can only be applied successfully to small regions which can be scanned rapidly. Otherwise appreciable changes may take place between the start and finish of a scan and the count will be unrepresentative. Consequently the only organ which has been regularly investigated with this method is the thyroid gland, for which scans can be repeated at intervals of four minutes or less. Robertson *et al* (1971) used it successfully to examine the handling of radioiodide and ^{99m}Tc pertechnetate by the thyroid gland. The method has now been largely superseded by the introduction of the gamma camera which has a faster response and more convenient methods of data analysis. More detailed information on scanners and their operation is given by Hine and Erikson (1974).

3.7.2 Gamma camera
The rectilinear scanner takes a finite time to build up an image since the

detector has to pass over all points in the radionuclide distribution. The gamma camera visualises its entire field of view simultaneously and continuously. This is a significant advantage for following the movement of radioactivity, especially rapid changes, e.g. those due to blood flow. The majority of gamma cameras in clinical use operate on the principle (see below) devised originally by H O Anger in 1956. A very small number of gamma cameras use an array of small crystals and operate on the autofluoroscope principle due to Bender and Blau (1963), use an image intensifier and a solid state localiser (Driard *et al* 1978) or are designed for use with the 511 keV annihilation photons from positron emitting radionuclides (Anger 1967). The latter three designs will not be discussed further here and the reader is referred to the original references. The principle of operation and the specification of perform-ance of Anger type gamma cameras have been described in detail in Horton (1978). Only an outline description of their operation will be given here for completeness.

A typical gamma camera with its control console is shown in figure 3.12 and a simplified sectional elevation of the detector assembly in figure 3.13. Gamma radiation from the radionuclide distribution in the patient reaches the large circular NaI(Tl) crystal through the multihole lead collimator. The crystal is typically 12 mm thick and has a field of view of 250 mm in standard cameras and 350–400 mm in large field cameras. As described in § 3.2.1, each incident γ-ray produces a scin-tillation whose photons reach a number of nearby photomultiplier tubes mounted in a regular pattern over the rear face of the crystal. Most cameras have 37 or 61 photomultiplier tubes. The amount of light reaching each photomultiplier tube, and therefore the amplitude of the output pulse, will be inversely related to its distance from the origin of the scintillation. The relative pulse heights from the photomultiplier tubes are compared by pulse arithmetic circuits to derive the position of the scintillation. This is given as a pair of x and y position signals, the amplitudes of the signals corresponding to the x and y coordinates of the scintillation about some nominal origin. The outputs of all the photomultiplier tubes are also summed to provide a z pulse whose amplitude corresponds to the total energy absorbed in the event causing the scintillation. This pulse undergoes pulse height analysis and if its amplitude lies in the window (see § 3.3.4) set for the radionuclide used, the x and y signals are allowed to pass to the image output devices. The simplest image output device is an oscilloscope with a Polaroid camera attachment, which is sited in the control console. The x and y position

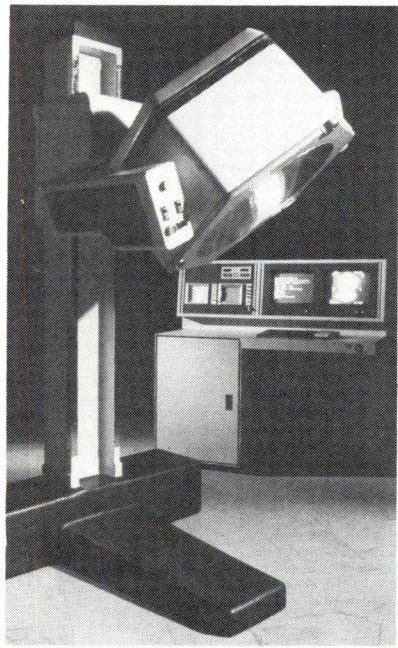

Figure 3.12 Gamma camera. A large field of view gamma camera with the detector (foreground) and the control console and image analysis computer (background). (Reproduced by kind permission of Scintag–Berthold Ltd.)

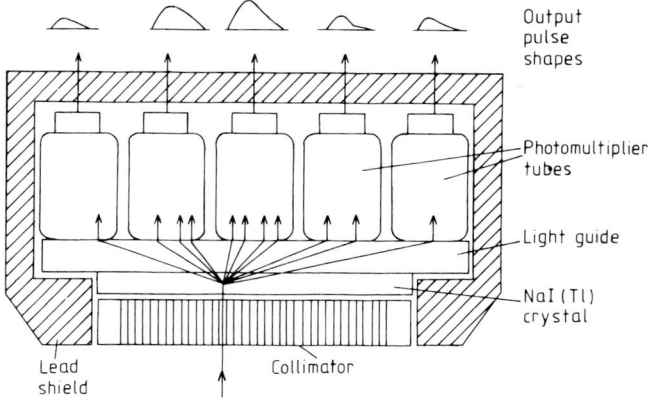

Figure 3.13 Gamma camera schematic. A sectional elevation showing the main components of the detector and the relationship of the pulse outputs to the position of the scintillation.

signals are used to shift the beam of the oscilloscope to a position corresponding to the original scintillation, which is then enabled to give a fine dot of short duration on the screen. The dots are recorded on Polaroid film for a preset time or number of counts. The film therefore integrates the events and a distribution of dots whose numbers reflect the radiation pattern reaching the crystal is obtained. Polaroid film is popular because it provides an immediate result, but suffers technically from having a limited grey scale response. Better quality images are obtained by using single sided radiographic film or 70 mm photographic film.

A number of different types of collimator are made for gamma cameras for different clinical purposes and can be changed readily and quickly during clinical work. The most common is the parallel hole type which transmits a pattern of radiation onto the crystal which corresponds directly to the radionuclide distribution in front of it. Diverging hole collimators, which have a virtual focus behind the camera, are used to diminish large organs, e.g. the lungs. Pin hole collimators and converging hole collimators with a virtual focus on the far side of the patient are used to provide magnified images. The former collimator has appreciable magnification and is used for small organs, e.g. the thyroid gland and the eyes, and for paediatric imaging. The latter collimator is used with large field cameras to provide better utilisation of the field of view for smaller organs, most often the brain. Due primarily to inverse square law effects, the sensitivity of diminishing and magnifying collimators changes from the centre to the edge of the field of view. Such collimators should therefore not be used for quantitative work without appropriate corrections, since different parts of the radionuclide distribution will be imaged with different sensitivities. It is always simpler to use a parallel hole collimator for which no such correction is required.

3.7.3 Image analysis

The images provided by the gamma camera on film can only be analysed qualitatively and subjectively. For quantitative analysis, the gamma camera is connected to a dedicated minicomputer or a microprocessor and the image is digitised into a square matrix of cells or elements (often termed 'pixels'). Matrices of 32×32, 63×64, 128×128 and 256×256 pixels are used—the first two principally for dynamic studies where the acquisition time is limited and the counts per pixel are low, and the latter two for static images where the counts per pixel and hence the accuracy (§ 3.8) can be made high by a suitably long acquisition time.

During data acquisition, the x and y position signals are digitised by analogue to digital convertors into one of the above binary ranges. The two digitised positions are then combined into the address of a location in the computer memory. In 'frame mode' acquisition, the content of the addressed location is increased by unity and the image is built up as a two dimensional histogram until either a preset time or preset total count is reached. A series of such frames can be acquired to show the movement of the tracer and stored on a magnetic disc. For studies with very high data acquisition rates, 'list mode' acquisition is employed in which the memory addresses together with timing marks are stored directly on a magnetic disc. This reduces the time taken to process events. The data are subsequently formed into frames with a specified preset count or time when acquisition is complete.

The stored images can be recalled for display and quantitative analysis. Images are usually displayed on a monochrome or colour television monitor, the former using a grey scale and the latter a colour scale (cf the scanner, see § 3.7.1) to display the different levels of pixel counts. Regions of interest (ROI) can be delineated on the image or summed image of choice using a light pen or a marker in the image controlled from a VDU and keyboard. The counts in the pixels in the ROI can be summed to provide a total count and the same ROI can be applied to a series of images to show the variation of activity with time. These counts can be expressed as a fraction of the dose by imaging a known aliquot of the dose under similar geometrical conditions to the patient and following the same analytical procedure.

Quantification using a gamma camera with image processor has the advantage over uptaké counting (§ 3.5) that it is applied accurately to the ROI selected from the image and not to the whole field of view. It is also possible to identify a background region indicative of tissue or vascular radioactivity; this is an important correction for the tissue overlying or underlying the organ of interest in many quantitative studies. A minor disadvantage with the gamma camera is that it cannot be used with radionuclides emitting gamma radiation with energies in excess of 400 keV. This is because of limits on shielding and collimator weight for the safe support of the detector assembly. Uptake counters with suitable collimation can be used with these radionuclides.

Care must be taken in the calculation of results from the camera when high activities are used, e.g. with bolus injections for cardiac studies. At low activities, the measured count rate is linearly related to activity, but at higher activities the observed count rate falls below that expected from extrapolation of the low count rate relationship, especially in the presence of scattering material. As a result, the count rates from those parts of a study where a high activity bolus is completely within the field of view tend to be underestimated.

Most image processors have routines for improving the display of static images. These are generally unnecessary for the analysis of dynamic studies except perhaps to provide a clearer image for the selection of ROI. Any routine which alters the numerical content of the acquired frames should be avoided as this may lead to scaling or rounding errors. In particular, frames should not be corrected for non-uniformities in the response of the camera. This is normally done by multiplying each pixel count by the corresponding correction factor, which is inversely related to sensitivity. These factors are obtained by imaging a uniform source of radioactivity and calculating the matrix of factors needed to yield a uniform pixel count. However Wicks and Blau (1979) and other workers have shown that non-uniformities in camera response are not usually the result of local variations in intrinsic sensitivity but are due to spatial distortions by which events crowd into some sites (giving a greater count density) and disperse from others (giving a lower count density). Because the distortions have a positional dependence, the corrections derived for a uniform source of radioactivity should only strictly be applied to a source of the same shape. To apply them to a radionuclide distribution having a different shape will result in errors, as pixels near the edges are incorrectly scaled up or down. The positional distortions will have no effect inside an ROI and little effect at the edges since the ROI covers a number of pixels.

3.8 Counting Statistics

3.8.1 Probability distributions

Radioactive decay is a random process and any count measurement will have a statistical uncertainty associated with it. This applies whether the count is determined from a sample *in vitro*, from an *in vivo* counter or from an ROI in an investigation using a gamma camera and image processor. The probability function P for repeat counts (r) for a given time interval is described by the Poisson distribution:

$$P(r, n) = \frac{e^{-n}n^r}{r!} \tag{3.1}$$

where n is the mean count determined from many such intervals. The

Poisson distribution may be applied to radioactive decay since the number of possible disintegrations is usually large and each disintegration is truly independent of all others. The probability of disintegration should also be small and constant and this is true in most circumstances, when the total period of measurement is small by comparison with the mean life of the radionuclide. The Poisson distribution is completely specified by the mean count, n, and it can be shown mathematically that the standard deviation, σ, is given by the expression:

$$\sigma = \sqrt{n}. \tag{3.2}$$

When n is small, the probability distribution $P(r, n)$ is very asymmetric, being skewed towards the low count values. As n increases the asymmetry becomes less marked and the shape approximates to that of the normal or Gaussian distribution with a mean n and a standard deviation $\sqrt{n}$. For $n = 10$, the normal distribution is a reasonable approximation if displaced towards the lower count values to take account of the asymmetry. For $n \geqslant 30$, asymmetry is negligible and the fit extremely close.

The fact that the standard deviation of the normal distribution is given by $\sqrt{n}$ is convenient for checking the stability of counting equipment. This also uses the property of the normal distribution that 95.4% of measurements lie within two standard deviations of the mean. A series of repeat counts of a constant activity can be made on a piece of equipment after installation or repair and the mean and standard deviation of the counts calculated. The number of counts outside the range $n \pm 2\sqrt{n}$ is then calculated and if this is close to 5%, the equipment is operating satisfactorily. If the number exceeds 5%, the counter is unstable and the cause should be found. These two features of the distribution also enable a quick check to be made that one count is significantly different from another, usually the background count. Each count (n_1 and n_2) is assumed to be part of a series for which there is a 95% chance that they lie within a range of four standard deviations $[(n_1 + n_2)/2]^{1/2}$. If they do lie within this range, they are probably not significantly different.

3.8.2   Effect of background count rate

Measurements of radioactivity require the subtraction of the background count rate, b, from that of the sample and background, a. Thus the net sample count rate is:

$$s = a - b. \tag{3.3}$$

Both a and b are subject to statistical variations and the uncertainty in the net count rate s will be a function of both. Suppose the sample is counted for time t_a and the background for t_b. Then the gross sample and background counts will be at_a and bt_b respectively. The standard deviations in the count rates will be:

$$\sigma_a = (at_a)^{1/2}/t_a = (a/t_a)^{1/2} \qquad \text{for the sample}$$

$$= \left(\frac{n}{t_a^2}\right)^{1/2} = \frac{\sqrt{n}}{t_a}$$

$$\sigma_b = (bt_b)^{1/2}/t_b = (b/t_b)^{1/2} \qquad \text{for the background}$$

and for the net sample count rate using the expression for the combination of errors:

$$\sigma_s = (\sigma_a^2 + \sigma_b^2)^{1/2} = (a/t_a + b/t_b)^{1/2}. \tag{3.4}$$

If the total time available for counting is $T(=t_a + t_b)$, then it can be shown that σ_s is a minimum when T is divided between the two measurements in proportion to the square of the individual count rates, i.e.

$$\frac{t_a}{t_b} = \left(\frac{a}{b}\right)^{1/2}. \tag{3.5}$$

3.8.3 Detection limit

To be detectable, a sample must give rise to a count rate which exceeds the statistical fluctuations of the background count rate. In general, the criterion used for minimum detectable activity is that it has a count rate equal to three standard deviations of the background count rate. If the background count rate is b measured over a time t_b, the standard deviation in the count will be $(bt_b)^{1/2}$ and in the count rate $(b/t_b)^{1/2}$. The minimum detectable activity is then $3(b/t_b)^{1/2}$ counts per unit time.

3.8.4 Optimisation of counting

For the measurement of samples containing large amounts of radioactivity, the background count rate is not important and the sample counts should be optimised by using the counting system with the greatest efficiency. If the measurements are to be done in a given time, expression (3.5) applies and a/b should be maximised.

For samples containing low activity, the background count rate is more important and Loevinger and Berman (1951) have proposed the use of a 'figure of merit' given by a^2/b. The shortest overall counting time is obtained when the figure of merit is a maximum.

4 Liquid Scintillation and Other Counting Techniques

4.1 Liquid Scintillation Counting

4.1.1 Introduction

Due to their limited range of detection, the application of beta emitting radionuclides is limited to investigations where the sampling of tissues or excreta is sufficient for diagnosis. Solid scintillation counting of samples is not possible since the radiation is absorbed by the walls of the container. Liquid scintillation counting, or internal scintillation counting as it is sometimes called, is now the method of choice for beta emitting radionuclides and a few radionuclides having low energy (less than 50 keV) x or gamma radiation. In clinical practice, the most common of the former radionuclides are ^3H and ^{14}C, the only long-lived radionuclides of the biologically important elements, hydrogen and carbon, and also ^{35}S and ^{32}P. The latter include ^{125}I and ^{55}Fe.

In liquid scintillation counting, the radioactive sample is intimately mixed with a liquid scintillator to ensure a minimum of absorbing material between the atoms of the radionuclide and the scintillator. This ideal spherical geometry for detection of the radiation results in high counting efficiencies, 50–60% for ^3H and 95% for ^{14}C. Both sample preparation and the computation of results are more complex than for gamma counting; preparation because the sample must be solubilised or homogenised to form a solution or suspension with the scintillator, and computation because individual counting efficiencies must be determined and corrections applied to each sample because of processes which reduce the intensity of the scintillations. These processes are generally termed 'quenching'. The whole technique is also commonly called 'beta counting'.

4.1.2 The scintillation process

The liquid scintillator comprises an aromatic *solvent* containing one or

60

more fluorescent aromatic *solutes* and in some instances other materials to aid the incorporation of the radioactive sample. The passage of a β-particle through the scintillator causes excitation and ionisation of the solvent molecules. These combine rapidly with free electrons to form excited solvent molecules. About 90% of these dissipate their energy thermally, but the remainder undergo internal conversion to the lowest excited singlet state. Through thermal excitation (Brownian motion) and excitation migration, this solvent excitation moves rapidly through the scintillator until it is transferred to a solute molecule. The solute is chosen so that its lowest excited singlet state is below that of the scintillator; if the concentration of the solute is sufficiently high the transfer of energy is virtually complete. The fluorescent efficiency of the solute is high (i.e. the ratio of photons emitted to excited molecules is virtually unity) and scintillation occurs. Common solvents are toluene, dioxane and xylene; solutes are discussed below.

A typical arrangement for the detection of the scintillation photons is shown in figure 4.1. The liquid scintillator, together with the radioactive sample, is placed in a 10 or 20 ml glass or polythene screw-capped vial. This is sited mechanically in a light guide in a light-tight chamber between two photomultiplier tubes. The tubes are sited close to the container for maximum efficiency of detection and the thickness of the light guide is minimised to reduce Cerenkov radiation. To avoid creating phosphorescence the vial should not touch the sides of the light guide on entry. The scintillation photons reaching the photocathodes of the photomultiplier tubes will give rise to electronic pulses as described in §3.2. These photons have a distribution of energies corresponding to the fluorescence spectrum of the solute and for optimum performance this should match the photoelectric spectral response of the photocathodes. The most common solute, 2,5 diphenyloxazole (PPO), has a spectral output centred around a wavelength of 370 nm, which is suited to modern quartz window bi-alkali photocathodes with a broad response. Older glass window photomultiplier tubes had an optimal response at longer wavelengths, around 430 nm, and a secondary solute was employed to shift the mean wavelength of the scintillations. This was commonly 1,4 di[2-(5-phenyloxazolyl)]benzene (POPOP), which has a lower energy excited singlet state than the primary solute (PPO) so that efficient solute–solute excitation transfer occurred. With modern equipment, the addition of a secondary solute is only necessary for opaque or coloured samples or for large volume samples in which self-absorption of the primary solute fluorescence occurs.

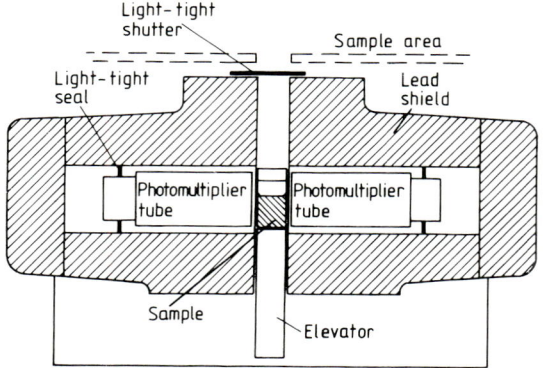

Figure 4.1   Detector for liquid scintillation counting.

With low energy radionuclides, e.g. ^{3}H, the number of photons emitted at the photocathodes of the photomultiplier tubes due to scintillations is comparable with the number due to thermionic emission (the dark current). Phosphorescence of the sample due to light exposure during preparation is another source of background photons. These two effects are nullified by operating the two photomultiplier tubes in coincidence. A typical schematic arrangement is shown in figure 4.2 with some of the electronic modules described in § 3.3. The outputs of the two photomultiplier tubes are summed to produce a pulse height spectrum closely related to the radiation energy spectrum. The outputs of the photomultiplier tubes also go to a coincidence circuit. This only allows a sum pulse to undergo pulse height analysis and be counted if the pulses from the two photomultiplier tubes occur within the resolving time of the coincidence circuit (typically 10 ns). Pulses arising from the photons from a single scintillation will meet this criterion and the sum pulse will be counted. A noise pulse from one photomultiplier tube will not be correlated in time with noise pulses from the other photomultiplier tube (except by chance) and will not usually be counted. Similarly, single photon events such as phosphorescence can only give rise to a pulse from one photomultiplier tube and will correlate only by chance and at high intensities. The introduction of coincidence counting has removed the need to operate liquid scintillation counters around 0°C to reduce the dark current. However operation at constant temperature is advantageous to ensure constancy of spectral output from the liquid scintillator.

In liquid scintillation counting, measurements of the background count rate are made using a vial containing liquid scintillator and no radioactive sample. Typically only one third of the rate measured is true background derived from cosmic rays, ambient radiation and shielding materials. The remainder is due equally to the glass vial and to light interactions between the two photomultiplier tubes. The former can be reduced by using potassium-free vials (to reduce their natural ^{40}K content), or polythene vials, although these have the disadvantage of being slightly permeable to toluene, a property which sets a limit to the accuracy of low count measurements.

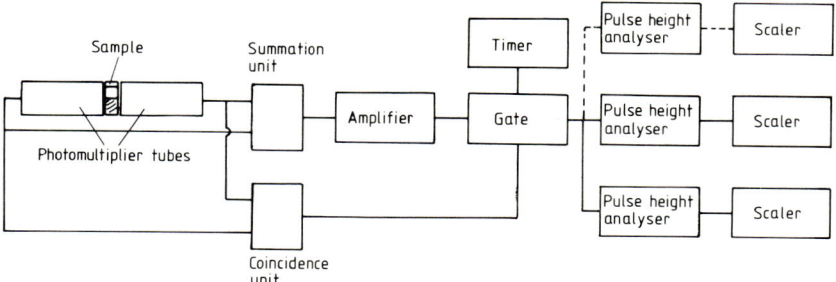

Figure 4.2 Electronic modules for liquid scintillation counting.

4.1.3 Sample preparation

Various methods of incorporating radioactive samples into liquid scintillators, depending primarily on the nature of the sample and its solubility in the scintillator, are available.

The simplest technique is dissolution in toluene. However, this has limited application. Steroids, lipids and fatty acids may be handled in this way.

Chemical solubilisation to facilitate dissolution in toluene scintillators is more common. The commonest solubilising agent is hydroxide of hyamine. This is however expensive and introduces quenching (see § 4.1.4) and chemiluminescence. The latter is particularly marked in alkaline preparations and the pH should be restored to neutrality by the addition of glacial acetic acid. This method is employed for proteins, amino acids, nucleic acids and for the trapping of $^{14}CO_2$ in expired air and from combusted samples (see later).

The solubility of aqueous samples is very limited in toluene scintillators. Generally, for homogeneous preparations, the more miscible dioxane scintillator is used. This comprises dioxane, PPO and naphthalene, the last behaving as a secondary solvent to transfer the excitation energy from the dioxane to the PPO. This scintillator will accept about 20% of its volume as water. It is widely used for counting blood, urine and tissues.

Emulsions provide a more efficient way of counting aqueous samples and can accept up to equal parts of water and scintillator. The liquid scintillator is prepared from toluene or p-xylene, PPO and Triton X-100, an emulsifying agent. Emulsions are more efficient than homogeneous scintillators like dioxane for two reasons. Firstly, the scintillator tends to concentrate in the oil-rich phase whilst the quenching agents and the salts tend to concentrate in the water-rich phase. As long as the water content is sufficiently low, the β-particle may escape from the aqueous phase, with its higher quenching action, to the oil-rich phase where scintillation can take place without quenching. Secondly, the scintillation process takes place in toluene or p-xylene, which are more efficient than dioxane. These emulsions are commercially available and many biological samples can be incorporated with little or no preparation.

Insoluble materials, especially white powders, can be counted successfully as suspensions. The sample is mixed with a thixotropic gel and the scintillator, usually toluene, added. This method has been applied to bone ash and bacteria.

For dry samples, especially those containing ^{14}C or ^{3}H, combustion is often used to prepare the sample for counting. The sample is pressed into a pellet and is ignited by a heating element under a stream of oxygen gas. $^{14}CO_2$ produced from ^{14}C labelled material is trapped in an alkaline solution, usually hydroxide of hyamine, and tritiated water (HTO) from tritiated material is collected by cooling. The contents of the traps are transferred to counting vials and the appropriate scintillator added. This method has the advantage that all samples are similar in composition and have similar counting efficiencies.

Cerenkov counting has been employed in liquid scintillation counters for radionuclides having energetic β-emission, e.g. ^{32}P. Cerenkov light is produced when the β-particle has a velocity greater than the velocity of light in the surrounding medium. There is an energy threshold, related to the refractive index of the medium, below which the effect does not take place. In practice, the most common medium is water, for which the threshold is 263 keV; this precludes its use for ^{14}C and ^{3}H. The advantages of the method are the ease of preparation of samples, avoidance of expensive scintillator and absence of quenching effects. However, because of the very directional light output and the poor match of the spectrum to the spectral response of most photomultiplier tubes, counting efficiencies are low. Methods of preparing samples are described in more detail by Turner (1971).

4.1.4 Quench correction

Since the total energy of each β-particle is absorbed within the scintillator, the observed pulse height spectrum should correspond closely to the β-particle energy spectrum. In practice impurities can reduce the intensity of the scintillations and hence the observed pulse heights by competing with the energy transfer from radiation to solute, in particular by dissipating the excitation energy of solvent and solute molecules. This is termed *impurity* or *chemical quenching*. Impurities clearly arise from the radioactive sample and its volume should obviously be kept to a minimum compatible with the counting accuracy desired. However a 20% loss of scintillation efficiency can also arise from dissolved oxygen, and scintillators should always be stored under a nitrogen atmosphere.

In addition, highly coloured or opaque materials in the sample will absorb some of the scintillation photons before they reach the photocathodes. This effect is termed *colour* or *optical quenching* and is particularly marked with yellow samples. The effect can be reduced by bleaching the sample before addition of the scintillator; addition of hydrogen peroxide is a common method.

The total effect of both types of quenching is to reduce the intensity of the scintillations and make them appear to arise from β-particles of lower energy. A downward shift in the pulse height spectrum is observed (figure 4.3). Quenching effects will vary from vial to vial because of the individual nature of the radioactive sample. The loss of low energy events due to the spectral shift and hence the lower counting efficiency must therefore be measured for each vial to calculate the true radioactive content. This procedure is called '*quench correction*' and a number of ways of doing it are described below.

Internal standardisation. The sample is counted with a single pulse height analyser and scaler. A constant activity of the same radionuclide, preferably in the same chemical form as the sample, is added to each sample (often called 'spiking') and the samples recounted. The additional count rate, D, due to the added radioactivity can be compared with that, S, due to the same activity in an unquenched sample to provide the efficiency of counting, n, i.e.

$$n = D/S.$$

True sample activity is then calculated by dividing the net sample count rate by n.

The added material must not alter the quenching of the sample and counting should take place as soon as possible after addition. With toluene scintillators, ^3H or ^{14}C toluene are often used. The added activity must be at least as great as the sample activity and should be sufficient for good statistical accuracy. The method is not popular because of the labour of accurately adding a constant activity to all the samples and the time spent in recounting the samples.

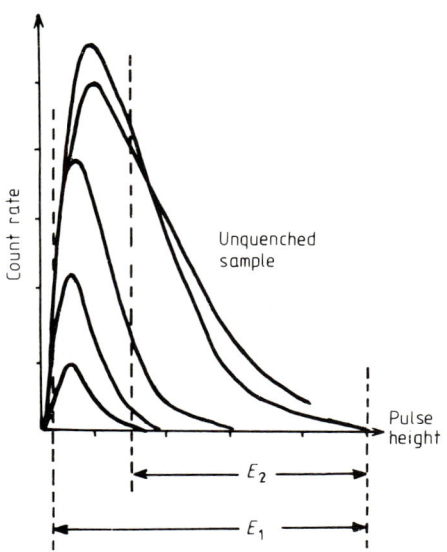

Figure 4.3 Quenched β-spectra. Typical window settings for quench correction using the sample channels ratio (SCR) are shown; SCR = E_1/E_2.

Sample channels ratio. This method exploits the shift in the pulse height spectrum with quenching and employs two channels of pulse height analysis. These are usually positioned as shown in figure 4.3, one covering the whole spectrum and the other the lower section. A series (between 5–10 vials) of quenched standards containing equal amounts of the radioactivity used in the study is prepared and increasing amounts of quenching agent are added, starting with none. This agent is commonly acetone, carbon tetrachloride or bromoform. The quenched standards are counted together with the samples under the same conditions. The sample channels ratio is the ratio of net counts in the low energy channel and in the wide energy channel. For each quenched standard the efficiency of counting is determined from the ratio of counts to the unquenched standard in the wide energy channel. These values are plotted against the sample channels ratio to produce a 'quench correction curve'. Typical quench correction curves for ^{3}H and ^{14}C are shown in figure 4.4(a). The range of quenching in the standards should cover the range expected in the samples. For each sample, the sample channels ratio is calculated and the corresponding efficiency found from the correction curve. The sample count in the wide energy channel is divided by the efficiency to yield the true count rate.

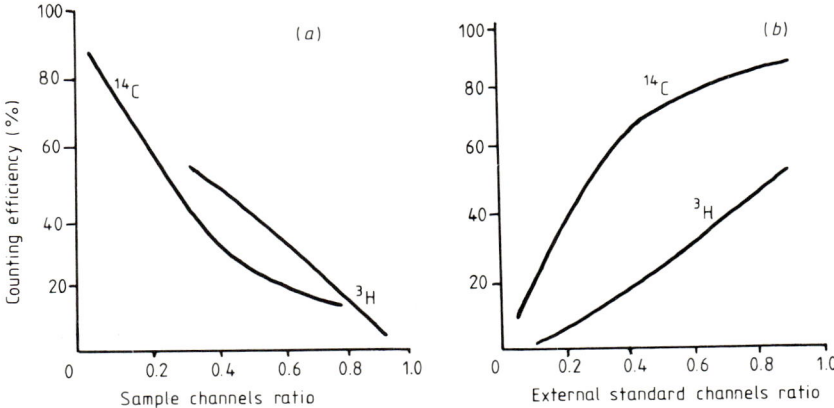

Figure 4.4 Quench correction curves for ^{3}H and ^{14}C using (a) sample channels ratio and (b) external standard channels ratio.

The sample channels ratio method is generally the most accurate method since it relates to the behaviour of the whole sample and it is independent of sample activity. However it clearly becomes inaccurate when the counts in the sample are low giving errors in the channels ratio. It is also limited to the range of discriminator settings which give good statistical results. Channel settings for heavily quenched samples may be unacceptable for lightly quenched samples where a steep correction curve may result and efficiency determinations become imprecise.

External standard channels ratio. An external standard is a gamma emitting source which is usually moved pneumatically into the vicinity of the counting vial after the sample counts have been determined. The gamma radiation produces Compton electrons in the sample and these give rise to a pulse height spectrum similar to the sample. ^{133}Ba, ^{137}Cs and ^{226}Ra sources have been used.

Originally the use of the external standard was introduced as a more convenient alternative to internal standardisation (see earlier) and the total count due to the standard was determined in a single channel. The efficiency of counting diminished with increased quenching. However the accuracy of the method was limited because it was sensitive to differences in sample volume, source decay and reproducibility of source position. These difficulties are overcome by using the external standard channels ratio.

After the sample has been counted in a channel set by the user, the external standard is brought close to the sample for 30 s and the induced counts measured in two factory-set channels matched to the spectrum of the particular standard source. The shift in the pulse height spectrum from the Compton electrons is correlated with the efficiency of counting of the sample in its own counting channel. The efficiency of counting in the sample channel and the channels ratio for the external standard are determined for a series of quenched samples and plotted to give a quench correction curve as described earlier. Typical external standard quench correction curves for ^{3}H and ^{14}C are shown in figure 4.4(b). For each sample, the channels ratio due to the external standard is used to derive an efficiency value from the curve which is divided into the sample count rate to give the true count rate.

This method is the most convenient for quench correction and has the advantage that it is independent of sample activity. In particular it can be used for greater accuracy when sample activity is low. However it is not necessarily the most accurate and problems can arise in its use with inhomogeneous samples.

4.1.5　Dual label counting

When counting two gamma emitting radionuclides with a solid scintillation detector, the pulse height spectrum of the high energy radionuclide will overlap that of the lower energy nuclide. A channel covering the photopeak of the lower energy nuclide will also contain a contribution from the spectrum of the higher energy nuclide. This contribution is easily determined by counting a sample containing only the higher energy nuclide and determining the net count rates C_1' and C_2' in lower and higher windows respectively.

Suppose

$$\alpha = C_1'/C_2'.$$

Then in a sample containing both radionuclides, the count rate C_2 due to the higher energy radionuclide can be determined unambiguously from the count in the higher energy window and the count rate C_1 due to the lower energy radionuclide alone in the lower window can be determined from the expression:

$$C_1 = C - \alpha C_2$$

where C is the measured count rate in the lower energy window.

In the liquid scintillation counting of a mixture of two radionuclides, the factor α relating the relative contributions to the two counting channels is not a constant, but varies with the degree of quenching. Removing the contribution of the higher energy radionuclide from the channel covering the lower energy radionuclide is therefore more complex. In the general case if there are two channels of pulse height analysis they can include contributions from both radionuclides. Suppose E_1 and E_2 are the counting efficiencies of the lower energy radionuclide in the low and high energy channels respectively and E_3 and E_4 are the efficiencies of the higher energy radionuclide in the low and high energy channels respectively. If A and B are the net total count rates in the low and high energy channels and X and Y are the activities of the low and high energy radionuclides then:

$$A = E_1X + E_3Y$$

and

$$B = E_2X + E_4Y.$$

Solving for X and Y:

$$X = \frac{AE_4 - BE_3}{E_1E_4 - E_2E_3}$$

and

$$Y = \frac{BE_1 - AE_2}{E_1E_4 - E_2E_3}.$$

To improve the separation of the two radionuclides and simplify the calculation of results, the threshold of the high energy channel is often set to exclude counts from the low energy radionuclide as shown in figure 4.5(a). In this situation, $E_2 = 0$ and the expressions for the activities become:

$$X = \frac{A - YE_3}{E_1}$$

and

$$Y = \frac{B}{E_4}.$$

This procedure can generally be used when the ratio of the maximum β-particle energies is three or greater. Dual label counting can only use

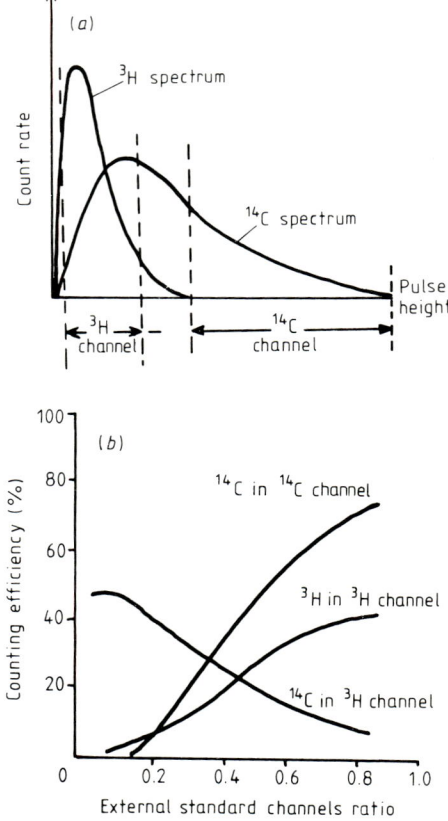

Figure 4.5 Dual label counting. (a) Typical window settings for simultaneous ^{3}H and ^{14}C counting and (b) quench correction curves using external standard channels ratio.

the external standard channels ratio method for quench correction since there are insufficient sample channels for the samples channel ratio method. Typical quench correction curves for the derivation of E_1, E_3 and E_4 (above) are shown in figure 4.5(*b*). The disadvantage of this method is that the channel settings are not optimal for counting efficiency. If the activity of the high energy nuclide is large, the subtraction of its contribution to the low energy channel can result in a low count rate with large errors. To reduce the contribution to the low energy channel, the window width of the low energy channel is often reduced, but this also reduces the efficiency of counting the lower energy radionuclide. Severe quenching also makes separation of the radionuclides difficult.

4.1.6   Commercial counters

The detection system employed in most sample counters is similar to that shown in figure 4.1 and the electronic arrangements similar to those shown in figure 4.2. Two channels of pulse height analysis are always provided for quench correction using the sample channels ratio method,

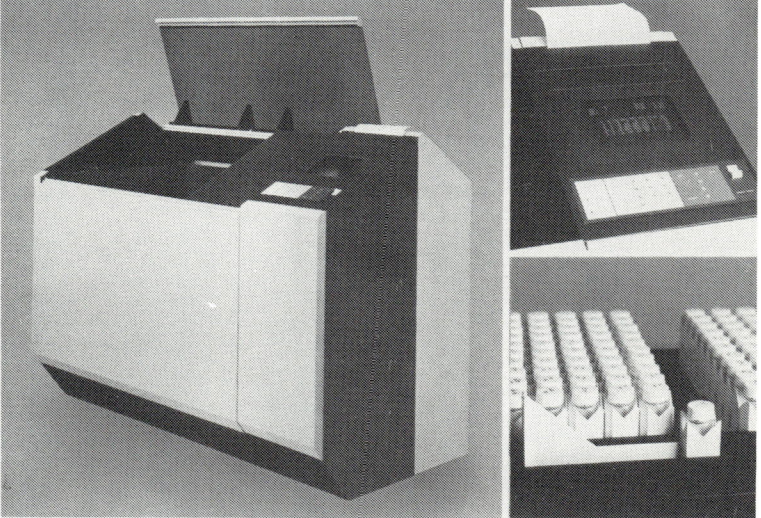

Figure 4.6   Automatic liquid scintillation counter. The counter is shown left with the access to the sample storage area open. The inset (upper right) shows the controls and the data display which use a microprocessor. The inset (lower right) shows sample vials in the tray storage system. (Reproduced by kind permission of Packard Instruments Ltd.)

and in some instruments three are provided. In the latter case one channel covers the whole pulse height spectrum and the other two cover the upper and lower halves to measure the spectral shift with quenching.

Sample transport arrangements are similar to automatic gamma systems. Continuous belt and tray systems are employed, both with the facility to select new counting conditions for each group of samples when a number of investigators use the counter. Counting conditions for the sample channels, the total counts and time and requirement for external standard counting are selected on the control panel. At the conclusion of each sample count according to the preset conditions, the sample identifier, counting time, counts in each sample channel and the counts due to the external standard in its factory-set channels are listed. In modern instruments, microprocessors are used to store the quench correction curve derived from a set of quenched standards and to correct each sample's net counts for the counting efficiency. An example of a modern liquid scintillation counter is shown in figure 4.6.

Fuller accounts of liquid scintillation counting are given by Birks (1964), Bransom (1970), Horrocks (1974) and Rapkin (1974).

4.2   Ionisation Chambers

Ionisation chambers are commonly used in nuclear medicine to measure

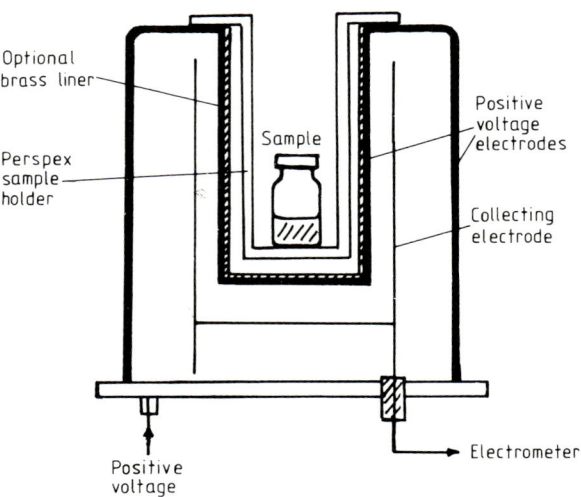

Figure 4.7   Ionisation chamber.

patient doses of radioactivity. Chambers are of the re-entrant type, with a cylindrical cross-section, like that shown schematically in figure 4.7. Radiation from the sample placed in the centre of the chamber produces ion pairs in the chamber gas. The chamber is operated at an applied voltage sufficient for the collection of all the ions formed without recombination or amplification occurring. Under these conditions, the current at the electrodes is proportional to the radiation intensity and hence the radioactivity of the sample. It is also proportional to the volume of gas in the chamber which is typically 10^5 mm^3. The current is measured with an electrometer and is typically 10^{-13} to 10^{-9} A. Chambers are normally unsealed and corrections should be made for temperature and pressure for changing gas volume.

For the measurement of doses of radioactivity, the ionisation chamber must either have a published reproducible calibration or be calibrated with known activities of each radionuclide. The 1383A chamber, commonly employed in the United Kingdom, comes into the first category. Developed by the National Physical Laboratory in conjunction with AERE, Harwell, and later produced by GEC–Elliott Automation Ltd, chambers of this type have calibration characteristics within ±1%. This chamber has a large geometry dependence and measurements must be made with the appropriate sample container. Because converting from a current reading to activity is not convenient, and is prone to error, commercial manufacturers have introduced instruments which

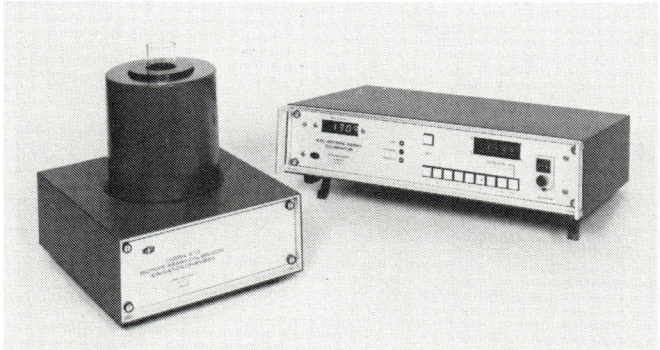

Figure 4.8 Radionuclide dose calibrator. The ionisation chamber is shown left with an electrometer having push button settings for the different radionuclides on the right. (Reproduced by kind permission of D A Pitman Ltd.)

read directly in becquerels or curies. These are often called 'dose calibrators' and an example is shown in figure 4.8. These are usually in the second category above and are calibrated by the manufacturer using known activities. Different factors are entered for each nuclide to take account of different radiation characteristics. Some instruments use a high pressure sealed chamber to improve sensitivity. Correction for non-linear response may be necessary at high activities. Chambers should be calibrated daily for consistency using a long-lived source such as ^{137}Cs or ^{226}Ra and at longer intervals for accuracy using standard solutions of the most frequently used radionuclides.

4.3 Radiochromatogram Scanning

As pointed out in §1.6.4, the presence of radiochemical impurities in a patient dose can lead to misleading and poor quality results and the irradiation of non-target organs; both are undesirable. The presence of such impurities is detected by standard chemical separation techniques, the most important being chromatography. With this technique, a test sample is deposited at one end of a strip of a suitable medium, e.g. paper, and the strip is then stood vertically with the end carrying the sample in a suitable liquid solvent. The solvent is allowed to diffuse up the strip until it reaches the far end. Materials of different molecular weight or moiety size move with the solvent to varying extents, the smallest usually staying close to the solvent front. As a result the compounds present become separated along the length of the strip. Paper and thin-layer chromatography (using glass strips coated with suitable absorbent materials) are commonly employed. The presence of the radioactivity in the radiopharmaceutical and its impurities enables them to be located on the strip and the proportions quantified.

A number of ways of measuring the radioactivity profile along the length of the strip are available. In the radiochromatogram scanner, the strip is scanned by a moving radiation detector, which is connected to a ratemeter system (§3.3.1) and a chart recorder. The distribution of radioactivity is recorded on the chart and the areas beneath peaks of activity may be integrated to give the activities of the components present. Geiger–Muller counters are used when ^{3}H and ^{14}C compounds are regularly analysed. For ^{3}H, a windowless counter whose volume is continuously flushed with an inert gas mixture is used. For beta emitting radionuclides with higher maximum energies in the range 150–500 keV, a thin end window counter is employed. The density of the window should be less than 10^{-7} g mm^{-3}. For gamma emitting radionuclides, a NaI(T1) detector having a very thin crystal is used.

With gamma emitting radionuclides, particularly ^{99m}Tc, a rapid method can be adopted if there is access to a gamma camera. The chromatograph strip can be placed on the face of the camera on a plastic sheet to prevent contamination. The image obtained shows immediately if there are impurities present and if the camera is connected to an image processor, the ROI facility (§ 3.7.3) can be used to quantify the amounts present.

Images of radiochromatograms are also obtained using 'spark chambers' or multi-wire proportional chambers. These comprise an anode grid at a positive potential of about 5 kV between two cathode grids. The wires in the two cathode grids are orthogonally orientated with respect to one another and the whole is contained in an inert gas atmosphere. An electron entering the chamber or one produced by ionisation by an incident γ-ray gives rise to an electron avalanche, similar to that produced in a Geiger–Muller tube. The electric field draws the ions apart and a pulse is induced in the wires in the cathodes around the position of the avalanche. Delay lines are used to derive X and Y position signals from the pulses on the orthogonal grids. These are processed in a similar way to those from the Anger gamma camera (§ 3.7.2) to provide a two-dimensional image. The resolution of such systems is good and it has the advantage (as has the gamma camera) that a number of chromatograms developed as parallel strips can be imaged simultaneously.

4.4   Semiconductor Detectors

Semiconductor detectors are often called solid state detectors since they behave as solid state analogues of the gas counter, electron–hole pairs being produced by the radiation instead of ion pairs. They comprise single crystals of high purity silicon or germanium having diffused p–n junctions and lithium drifted p–i–n structures. They have the advantages of a very linear response to energy, excellent energy resolution and a fast response and can be made very small. The energy resolution is typically four times better than a NaI(Tl) detector, but sensitivity is rather poorer. Their principles of operation and operating characteristics are described in greater detail by Friedland and Zatzick (1967).

The most common applications of semiconductor detectors in nuclear medicine are γ-ray spectroscopy and the measurement of tissue concentrations of radioactivity. The former is used to identify and quantify radionuclides from their energies of radiation and may be used in the measurement of radionuclidic purity. Spectroscopy is usually performed with a lithium drifted germanium (GeLi) detector connected to a multichannel analyser (§ 3.3.7). Tissue concentrations of radionuclides are sometimes measured with silicon detectors with a small sensitive volume, so that they can be introduced with minimal disturbance of the tissue. The uptake of ^{32}P in cerebral and ocular tumours has been measured in this way and the clearance of ^{85}Kr from the eye has been monitored to measure ocular blood flow.

5 Measurement of Volume and Mass

5.1 The Dilution Principle

The volume of distribution or the mass of a substance within the body can sometimes be measured by application of the isotope dilution principle. In theory the principle is very simple. A measured quantity of radioactivity is added to the system under investigation and allowed to come to equilibrium throughout it. A representative sample of known volume or mass is then taken. From the measured concentration, or specific activity, the volume or mass with which the radioactivity has been diluted is calculated. This principle is not new. Colorimetric and chemical methods have previously been employed since at least 1915. Radioisotopes were first used in 1934 when Hevesy used deuterium to measure total body water (see §5.4.1). In practice, application of the principle is rather more complex than the theory. This is because volumes of distribution may not have precise boundaries and/or equilibrium concentrations may be difficult to achieve. These practical limitations to the accuracy and interpretation of results will become clearer in the subsequent sections by reference to particular examples. In spite of these limitations, the dilution principle is a powerful and convenient tool for measuring irregular distributions of volume or mass.

The distribution of the radioactive tracer at equilibrium can take two forms as shown in figure 5.1. In the simpler situation shown in figure 5.1(a), the tracer is distributed uniformly throughout the volume of the system. This has been referred to as a 'well stirred' system. Suppose a given activity A in a volume v is added to this system, which has a volume V. The equilibrium concentration c throughout the system will then be given by the expression:

$$c = A/(V + v) \qquad (5.1)$$

and this can be measured by counting. In practice v is much less than V and expression (5.1) reduces to:

$$c = A/V. \qquad (5.2)$$

The volume of distribution is then given by the equation:

$$V = A/c. \qquad (5.3)$$

Physiological volumes measured using this technique include plasma and blood volume (§5.2 and §5.3).

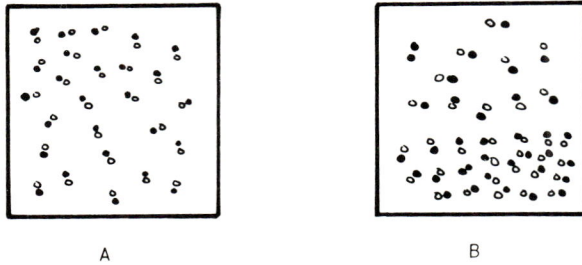

A   B

Figure 5.1   Evenly and unevenly distributed systems. In system A, the radioactivity (open circles) is distributed evenly and the concentration is uniform. In system B, the radioactivity is distributed with the same non-uniformity as the stable material (full circles); the specific activity is uniform. System A is termed a 'well stirred' system.

In the more complex situation shown in figure 5.1(*b*), the tracer is distributed in the same fashion as the identical non-radioactive material in the system, which may not be distributed evenly. In this situation, the specific activity of the tracer is constant throughout the system at equilibrium. Suppose the tracer has an activity A and a mass m. The specific activity of the tracer is then given by:

$$s = A/m. \qquad (5.4)$$

If the tracer is mixed with the inactive substance in the body of mass M, the new specific activity which can be measured from a sample is given by:

$$s^1 = A/(M + m). \qquad (5.5)$$

In practice, the tracer will have a high specific activity so that its addition does not perturb the system under investigation, and m will be very small. Equation (5.5) then reduces to:

$$s^1 = A/M. \qquad (5.6)$$

The mass of substance present is then given by:

$$M = A/s^1. \qquad (5.7)$$

The great advantage of this technique is that it is independent of sample volume. Since it is not restricted in its application to 'well stirred' systems, it enjoys a wider application than the measurement of volume described earlier. It is, however, more complex in practice, since it requires the determination of the amount of stable material present in addition to the measurement of radioactivity. Physiological masses commonly measured with this technique are those of exchangeable sodium and potassium (§5.4).

These concepts of mass and volume have been discussed elegantly and at greater length by Bergner (1967).

5.2 Plasma Volume

The circulating blood volume has an irregular but well defined boundary and is therefore well suited to measurement by the dilution principle. However direct measurement of the blood volume is not possible because of the very different natures of its principal components, plasma and red cells. The volumes of plasma and red cells have to be measured separately and added to give the blood volume. Measurement of the former is described in this section and the latter in the following section.

The tracer added to the plasma must remain within the plasma volume for equation (5.3) to apply, at least until such time as an equilibrium distribution is reached. All methods of measuring plasma volume use labelled plasma proteins, usually human serum albumin (HSA). Initially blue dye was used as the label, but radioiodine is now standard. Commercial preparations labelled with ^{131}I or ^{125}I are available. The latter is generally preferred for two reasons. Firstly, its 60 day half-life provides a product with a longer shelf life than is possible with the eight day half-life of ^{131}I. Secondly, the low energy (30 keV) radiation from ^{125}I causes less radiolytic decomposition than the more energetic gamma and beta radiation from ^{131}I. In particular this results in a higher radiochemical purity with less than 5% of the total radioactivity present as iodine or iodide. The usual dose of ^{125}I–HSA or ^{131}I–HSA is 185 kBq (5 µCi). The use of ^{51}Cr–HSA has also been reported but the results of measurements seem variable. This has been attributed to varying permanence of the label.

The details of the technique are as follows. The sterile dose is provided in 5–10 ml physiological saline solution. A known volume d (typically 5 ml) is injected intravenously into an arm vein and the remainder of the dose is kept for the preparation of a 'standard' whose activity is related to that of the given dose. A known volume s (typically 1 ml) of the remainder is diluted to a volume V (typically 100–500 ml) in a volumetric flask and a small volume v (the standard) is taken for counting in a well counter and scaler/timer (see §3.4.1). If a net count rate r_s is measured, the activity A administered is given by:

$$A = \frac{kd}{sv} V r_s \qquad (5.8)$$

where k is a constant relating activity to count rate and will be dependent on counting geometry, etc. Between 10 and 15 minutes after dose administration, a venous blood sample (typically 5–10 ml) is taken into anticoagulant from the opposite arm to that used for injection. (This avoids contamination of the sample by any residual activity at the injection site.) The sample is centrifuged and a known volume b of the supernatant plasma is taken for counting. If the measured net count rate due to the plasma sample is r_b, the activity concentration c is given by:

$$c = k \frac{r_b}{b} \qquad (5.9)$$

where k has the same significance as before. During the 10 minute interval between dose administration and taking the venous sample, some diffusion of the albumin into extravascular space takes place. This results in a lower concentration than that expected due to dilution alone and it is standard practice to apply a correction factor of 1.015 for a sample taken at 10 minutes. The true concentration is, therefore, given by:

$$c = 1.015 \, k \frac{r_b}{b}.$$

Plasma volume P may then be calculated using expression (5.3), i.e.

$$P = \frac{dVb}{1.015 sv} \frac{r_s}{r_b}. \qquad (5.10)$$

The use of a standard which is related by volume to the dose avoids absolute measures of radioactivity of the dose and the sample with the introduction of large errors. The activity for the standard needs to be diluted before counting to avoid measurements with a high count rate and appreciable dead-time losses.

In certain clinical circumstances, it may be necessary to improve the accuracy of determination of plasma volume. This can be done in two ways which add to the technical complexity of the technique. Firstly the accuracy of the relationship between the dose and the activity taken for the standard can be improved by weighing each in a syringe instead of measuring their volumes. The respective weights then replace s and d in equation (5.8). Secondly the correction factor 1.015 for losses from the plasma volume, whilst generally applicable, will vary in individual patients and losses can range up to 10% per hour in sick patients. To take account of individual loss rates, venous samples may be taken at 10, 20 and 30 minutes after dose administration and the activities per ml plasma measured. The loss of tracer is usually exponential and if the three count rates are plotted against time on a log–linear graph, a straight line can be fitted. If this line is extrapolated back to the ordinate (cf figure 6.2), the intercept gives the count rate that would have been observed at the time of injection if there were no diffusion losses present. This count rate may be substituted for r_b in equation (5.10) and the factor 1.015 omitted.

The technique can be extended to measure blood volume by using the packed cell volume or haematocrit. The haematocrit, h, is the volume ratio of red cells to whole blood. The blood volume B is therefore related to the plasma volume by the equation:

$$B = P/(1 - h). \qquad (5.11)$$

Haematocrit is measured by putting a sample of venous blood with anticoagulant into a fine bore tube sealed at one end and typically 100 mm long (often termed a Wintrobe tube). This is centrifuged at 1500g for 30 min after which the packed red cell length can be read as a fraction of the total length of the sample in the tube. This fraction h^1 is termed the uncorrected packed cell volume and must be corrected for two effects. Firstly, centrifuging at 1500g for 30 min is not sufficient to expel all the plasma from between the red cells and a correction factor of 0.96 must be used to allow for the effect of trapped plasma. Secondly, it has been found that there is a greater number of red cells per unit volume of blood in the venous blood vessels than in the body generally. The factor relating the whole body haematocrit to the venous haematocrit is 0.91. The corrected packed cell volume is therefore given by:

$$h = h^1 \times 0.91 \times 0.96. \qquad (5.12)$$

This is used in expression (5.11).

In some clinical conditions, this method does not provide accurate measures of blood volume. This occurs most often in patients with polycythemia, who have high haematocrits. It is possible in these patients that the ratio of 0.91 between whole body and venous haematocrits no longer applies. In this situation, the red cell mass must be measured separately and directly (as described in the next section) and added to the plasma volume to give the blood volume. Hobbs (1967) has described the measurement of plasma and blood volumes in greater detail.

5.3   Red Cell Volume

The determination of red cell volume must use labelled red cells since no other tracer form can mimic their discrete behaviour. The radioactive labels most commonly used nowadays are ^{51}Cr and ^{99m}Tc. ^{32}P is occasionally employed, especially for research, e.g. Bentley *et al* (1974), since as $DF^{32}P$ (di-isopropylfluorophosphonate) it is very strongly bound to the red cells. However, for routine purposes it has the disadvantage of requiring the samples to be assayed by liquid scintillation counting.

The labelling of red cells with ^{51}Cr is based on the technique of Gray and Sterling (1950) and is important since it is also used for red cell survival studies (§6.6.3). In this technique 20 ml of fresh venous blood in heparin anticoagulant is added to a sterile vial containing 1.85 MBq (50 μCi) of sodium (^{51}Cr) chromate in 2 ml physiological saline solution. The specific activity of the sodium chromate must be high (greater than 3.7 MBq $μg^{-1}$ or greater than 100 μCi $μg^{-1}$) to avoid poisoning the red cells. After gentle agitation to mix the cells and the radioactivity, the mixture is left in the dark at room temperature for 30 minutes to 'incubate'. During this period the chromate ions (CrO_4^{2-}) diffuse into the red cells where they are reduced to the chromic form (Cr^{3+}) and become attached to haemoglobin. Before re-injection of the labelled cells, any ^{51}Cr external to the cells must be removed. Otherwise, after injection, this activity may label plasma proteins *in vivo* and give an erroneously high activity in the equilibrium sample. The extraneous ^{51}Cr activity is removed by a simple washing procedure. The labelled red cells are gently centrifuged (500*g* for 5 min) and the supernatant plasma carefully removed and aseptically replaced with an equal volume of sterile physiological saline solution. The mixture is then gently agitated and the procedure repeated twice more before the cells are ready for injection. A number of variants of this technique exist. The two most important are (a) the use of other anticoagulants instead of heparin, for example acid citrate dextrose is said to provide energy for the red cells during the lengthy labelling procedure, and (b) incubation at body temperature instead of room temperature. In the author's experience, these differences have little effect on the efficiency of labelling of the red cells and their subsequent survival.

The labelling of red cells with ^{99m}Tc has been described in one of its most convenient forms by Eckelman *et al* (1971). Ten ml of fresh venous blood is added to a sterile vial containing 2 ml sodium citrate (2.8%) solution. After gentle agitation, the vial is centrifuged at 100*g* for 5 min and the supernatant removed. 0.5 ml of fresh stannous chloride solution $(4.4 \text{ mol} \, l^{-1})$ is then added to the red cells and the mixture incubated at 37°C for 5 min. During this period the stannous chloride enters the red cells. 1–3 ml of sodium (^{99m}Tc) pertechnetate solution with an activity of about 37 MBq (1 mCi) is then added and the incubation continued for a further 5 min. The stannous ions reduce the pertechnetate (TcO_4^-) to a more reactive form of technetium-99m which binds to the red cells. Physiological saline solution is then added to re-suspend the red cells and a two-stage washing procedure similar to that described for ^{51}Cr red cells is followed. The quantity of stannous chloride must be controlled because the red cells are damaged if excess quantities are used.

The measurement of red cell volume is similar to the measurement of plasma volume. A known volume of labelled cells is injected intra-venously and a small aliquot of the dose is kept for the preparation of a standard. This is added to a volumetric flask and the volume made up with water to which a small amount of saponin has been added to lyse the red cells and ensure a uniform distribution of radioactivity. The equilibrium venous sample is taken 10–15 min after dose administration. A known volume is taken for counting in a well counter together with the standard, and a Wintrobe tube is filled for the determination of haematocrit. In taking these samples, the blood must be thoroughly mixed to avoid errors due to the sedimentation of the red cells. Alter-natively, after taking the sample for haematocrit determination, the remainder of the sample may be lysed with saponin to ensure a rep-resentative sample for counting. The blood volume *B* is then given (cf equation (5.10)) by:

$$B = \frac{dVb}{sv} \frac{r_s}{r_b} \tag{5.13}$$

where *d* is the volume of activity injected, *s* is the volume of activity retained for the standard, *V* is the dilution volume of the standard, *v* is the volume of the standard taken for counting, b is the volume of the whole blood sample taken for counting, r_s is the net count rate due to the standard sample and r_b is the net count rate due to the whole blood sample. The red cell volume is then given by:

$$C = Bh \qquad\qquad (5.14)$$

where h is the haematocrit corrected for trapped plasma, i.e. $h = h^1 \times 0.96$ where h^1 is the uncorrected haematocrit obtained by the same procedure as described earlier. No correction is necessary in this case for the difference in venous and whole body haematocrits, as one is measuring the dilution of labelled red cells with red cells.

Simultaneous measurements of plasma volume using ^{125}I–HSA and red cell volume using ^{99m}Tc or ^{51}Cr–red cells can be performed. Account must be taken of the contribution of the higher energy nuclide (^{99m}Tc or ^{51}Cr) to the counts in the window set for the lower energy nuclide (^{125}I) as described in §4.1.5.

5.4   Electrolytes and Body Water

The maintenance of normal electrolyte levels in body tissues and fluids is important for good health and is the result of many interacting active processes. Gross disturbance of body fluid and electrolyte patterns leads to widespread impairment of tissue and organ function. Conversely, many disease conditions and injuries alter the distribution of fluids and electrolytes. In routine clinical practice, disordered electrolyte metabolism usually becomes apparent through abnormal serum electrolyte levels. Whilst these indicate an abnormality, they rarely identify the cause, and may be due to a variety of reasons. The understanding of electrolyte disorders has been greatly improved in recent years by measurements of exchangeable sodium, exchangeable potassium, total body water and extracellular fluid volume in healthy and disease states. The difference between total body water and extracellular fluid volume also gives a measure of intracellular fluid volume, for which no direct measurement is possible. The above measurements are all applications of the dilution principle.

5.4.1   Measurement of total body water
Body water comprises 60% of body weight, 40% being intracellular and 20% being extracellular. The tracer used for the measurement of body

water by the dilution principle is tritiated water, ^{3}HOH. The normal dose is 18.5 MBq (500 μCi), which may be given orally or intravenously. Equilibrium is fairly rapid and a venous sample for estimation of the plasma concentration of tritium may be taken two to three hours after dose administration. The excretion of the tracer over this period is small and usually ignored. The tritium content of the plasma sample and a diluted aliquot of the dose are counted by liquid scintillation counting (§4.1). The total body water W is given by:

$$W = \frac{dVb}{sv} \frac{r_s}{r_b} \tag{5.15}$$

where d is the volume administered, s is the volume taken for the standard, V is the dilution volume of the standard, v is the volume of the diluted standard counted, b is the volume of the plasma sample, r_s is the net standard count rate and r_b is the net sample count rate.

Alternatively, total body water can be determined from a urine sample taken four to five hours after dose administration. The urinary bladder should have been emptied some 30 minutes previously to ensure that the bladder concentration is representative of the equilibrium concentration. The patient should also have nothing to drink until after the sample is taken so that the equilibrium concentration is maintained.

Antipyrene labelled with ^{131}I or ^{125}I has also been used to measure total body water. Whilst this agent brings the practical advantages of gamma counting techniques, it is excreted fairly rapidly. Account must be taken of this loss by counting sequential plasma samples (as described in §5.2) over a period of four to six hours.

5.4.2 Measurement of extracellular water

Extracellular water is that part of body water which is not inside cells. It comprises the plasma, interstitial water, water in connective tissues, cerebro-spinal fluid, water in the gastro-intestinal tract and in the aqueous humour of the eye. The ideal tracer must therefore diffuse readily into all these tissues and must not be retained in the cellular barriers to the last three entities. Tissue analysis shows that 90% of body chloride is extracellular, the remainder being largely in red cells. A radioisotope of chlorine would therefore provide a suitable tracer. However, the isotopes of chlorine are not convenient for clinical use; the only gamma emitting isotope ^{38}Cl has a 37 minute half-life which is too short for equilibrium studies and the long lived ^{36}Cl is a beta emitter. Measurements are therefore usually performed with the convenient gamma emitting isotopes, ^{77}Br or ^{82}Br, of bromine which has a similar chemical behaviour to chlorine.

A dose of 0.93 MBq (25 µCi) of ^{77}Br or ^{82}Br as potassium bromide is given orally. Equilibration takes place within four to six hours in the normal individual and the plasma concentration remains appreciably constant over at least 48 hours. A plasma sample is normally taken at 24 hours and counted, together with a diluted aliquot of the dose, in a well scintillation counter. The bromide space is then determined using expression (5.15). The result is multiplied by 0.9 to give the extracellular fluid volume. As noted above, this may be subtracted from the total body water to give the intracellular fluid volume.

5.4.3   Measurement of exchangeable sodium and potassium

In the measurement of the masses of sodium and potassium, equilibrium is not approached until 24–48 hours after dose administration. During this period considerable amounts of the tracer will be excreted and account must be taken of individual losses by direct measurement; to assume a standard rate of loss would lead to inaccuracies. In these circumstances, the distribution volume V is given by:

$$V = \frac{\text{amount of tracer retained}}{\text{concentration of tracer in plasma}}.$$

The term volume is unreal when applied to electrolytes since the tracer will be distributed throughout the potassium in the body. Instead the terms *potassium* or *sodium space* are used to indicate the volume of plasma which would contain the tracer at the plasma concentration. This concept is useful in comparing different disease states.

The mass of electrolyte M will be given by:

$$M = \frac{\text{amount of tracer retained}}{\text{specific activity of tracer in plasma}}.$$

This expression assumes that the tracer is in equilibrium with the stable electrolyte throughout the body. In practice, this is difficult to achieve within reasonable time. Equilibration of the tracer with electrolyte in bone can take weeks due to the slow diffusion processes involved. The concept of *exchangeable* mass is therefore used, meaning that part of the body's electrolyte mass which is readily able to exchange tracer with the plasma. True equilibrium is not reached therefore—just a stage when the plasma concentration is only changing slowly. This takes 12 hours in normal subjects and more than 24 hours in patients with fluid disorders, e.g. oedema. It is important always to give the duration of the estimation, e.g. 24 or 48 hours. The mass of exchangeable sodium is typically 70% of total body sodium.

To measure exchangeable sodium, 1.11 MBq (30 μCi) of ^{24}Na (half-life 15 hours) as sodium chloride is given orally. Urine is collected for the next 24 hours and at the end of this period, a 10 ml venous sample is taken. A second 24 hour collection is started and at the end of this period a second 10 ml venous sample is taken. The venous samples are centrifuged soon after being taken. Part is used for gamma counting of the ^{24}Na, together with an aliquot of the diluted dose, and part is used for estimation of the stable sodium concentration (mmol l^{-1}) using flame photometry. Samples of each urine collection are also taken for counting to estimate the activity excreted. Estimations of the sodium space and exchangeable sodium are made at 24 and 48 hours as follows.

The amount of tracer excreted, E, is given by:

$$E = kV_1u_1 \qquad \text{at 24 hours}$$

and

$$E = k(V_1u_1 + V_2u_2) \qquad \text{at 48 hours,}$$

where k is the factor relating count rate to radioactivity, V_1 and V_2 are the total urine volumes for the first and second 24 hour periods respectively, and u_1 and u_2 are the net count rate per ml for a representative sample taken for each collection. The activity retained R is therefore

$$R = kr_s\frac{Vd}{vs} - E$$

where k, r_s, V, v, d and s have the same significance as before. The sodium space S is then given by:

$$S = \frac{r_s(Vd/vs) - E}{r_p}$$

where r_p is the net count rate per ml plasma at 24 or 48 hours as appropriate. The total exchangeable sodium Na_e is given by:

$$Na_e = Sn$$

where n is the stable sodium concentration (mmol l^{-1}).

Alternatively the estimations can be based on urine samples. Spot urine collections are then taken from 23–24 hours and from 47–48 hours after dose administration. Small samples are taken for gamma counting and estimations of stable sodium content, and the remainder added to the appropriate 23 hour urine collection. The above expressions are again used but with the values for the net count rate and the stable sodium per ml urine substituted for r_p and n respectively.

Exchangeable potassium is measured in an identical way using 1.11 MBq (30 μCi) of ^{42}K (half-life 12.4 h) or 0.74 MBq (20 μCi) of ^{43}K (half-life 22 h). Potassium takes longer to equilibrate than sodium, because it has to cross cell membranes. Equilibrium is substantially complete in normal states in 20 hours, but takes longer in disease states. Equilibrium with potassium in the red cells and the brain takes over 60 hours. Comparisons of exchangeable potassium with estimates of total body potassium based on its natural ^{40}K content suggest that it under-estimates body potassium by 15%.

5.4.4   Simultaneous studies

It is convenient and often important to carry out simultaneous estimates of the fluid volumes and electrolyte masses to arrive at a complete understanding of the patient's condition. The measurement of total body water, extracellular fluid volume (ECF) and exchangeable sodium and potassium may all be performed at the same time. To do this ^{77}Br must be used for the measurement of ECF and ^{43}K for the measurement of exchangeable potassium. The gamma spectra of these two isotopes and ^{24}Na are then separable and their activities can be measured on an automatic well scintillation counter having three spectrometers. This technique has been described by Davies and Robertson (1973). Because of their differing rates of decay, the doses of ^{24}Na, ^{43}K and ^{77}Br used in this situation are 0.37 MBq (10 μCi), 0.74 MBq (20 μCi) and 0.18 MBq (5 μCi) respectively so that the sample activities at the time of counting are approximately equal. This is necessary to ensure accurate results after correcting for the contribution from the higher energy nuclide(s) to the lower energy channel(s). The tritium content of the samples is assayed by liquid scintillation counting at a later time when the activity due to the three gamma emitting isotopes has decayed to a negligible level.

6 Measurement of Absorption, Clearance and Uptake

6.1 Introduction

The metabolism of many compounds have a number of processes in common. These include:

(a) absorption from the gastro-intestinal tract into the blood stream,
(b) clearance from the blood,
(c) concentration (or uptake) by individual organs or excretion by the kidneys, and
(d) metabolism by the organ concerned with the release of metabolites, which may be metabolised further or excreted.

The complete description of these processes, together with measurement of the mass or volume of each part, is the aim of system modelling. This is the subject of Chapter 8. However, for clinical diagnosis, a complete description of the performance of the system is often unnecessary because sufficient information can be obtained from the observation of one particularly salient feature. A complete description is also often difficult to obtain because of the lengthy and invasive measurements required. Measurement of a single parameter is more easily tolerated by the patient and is more acceptable for prompt clinical management of the patient's condition. Common instances where radionuclides are used for isolated measurements of absorption, blood clearance or organ uptake, together with clinical examples, are examined further in the following sections.

6.2 Absorption Measurements

6.2.1 Introduction
The absorption of an orally administered radiopharmaceutical by the

gastro-intestinal tract can be measured, depending on its metabolism, by measuring the radioactivity

(a) in the blood,
(b) retained in the whole body,
(c) excreted in faeces,
(d) excreted in urine, or
(e) exhaled in the breath.

Measurements using each of these techniques are described below.

6.2.2   Measurement of tracer levels in the blood

A rapid but approximate measure of absorption is to measure the concentration of radioactivity in the blood. This requires counting of plasma from a venous blood sample, together with a diluted aliquot of the given dose, in a well scintillation counter. The plasma level may then be expressed as a percentage of dose per ml plasma or the fractional dose per ml plasma; higher levels than the normal range of values indicate hyperabsorption and lower values indicate malabsorption. Measurements are standardised to a particular interval after dose administration; this is often one hour.

The disadvantage of this technique is that it takes no account of the elimination of the radioactivity from the blood, which will vary from individual to individual. The rate at which the tracer is eliminated depends on the state of the body. For example, if body stores of the tracer are low, it will be eliminated rapidly, giving rise to a low plasma level and an impression that little has been absorbed. Conversely, if body stores are replete, blood clearance may be slow, giving rise to high plasma levels and an impression of good absorption. As a result of differences in body condition, a wide scatter of results can be observed which may be too imprecise for clinical diagnosis of disordered absorption.

Accurate results can clearly only be obtained if the elimination of the tracer from the blood is measured in each individual. This involves a double tracer technique, in which one radiopharmaceutical is given orally and the other intravenously immediately following. Clearly both tracers must have the same form in the blood for identical clearance rates. Since both tracers are present in the plasma samples, they must be separable for counting. Standards related by volume to the two given doses are also counted so that the activity of each tracer can be expressed as a percentage of fractional dose per ml plasma. These values are plotted against sample time; typical variations for oral and intravenous doses of calcium are given in figure 6.1. The clearance curve for the intravenous dose indicates how the absorbed tracer is eliminated from the blood following its absorption. The plasma concentration of the oral dose will therefore be the summation of many such clearance curves whose initial height will be given by the quantity of tracer absorbed with time. This is illustrated in figure 6.2.

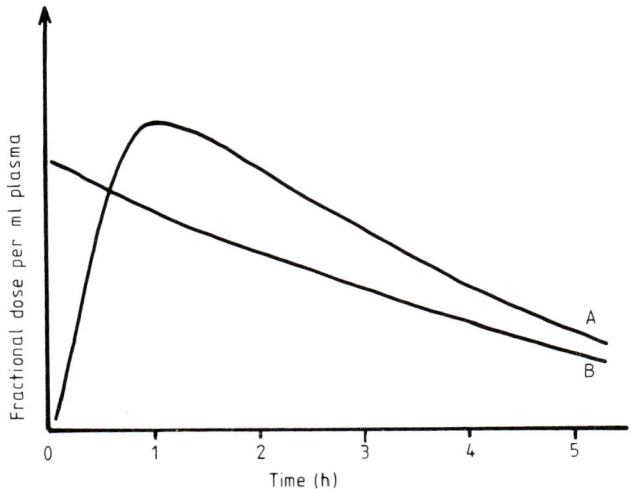

Figure 6.1   Intravenous clearance of calcium. For an oral dose (A) and an intravenous dose (B).

The process that is occurring is one of convolution. To achieve the variation in plasma concentration, $o(t)$, of the oral dose, the plasma variation of the intravenous tracer, $i(t)$, is convoluted with the curve $a(t)$ describing the fraction of the oral dose absorbed with time, i.e.

$$o(t) = \int_0^\infty a(t - \tau) i(\tau) \, d\tau. \tag{6.1}$$

The expression $a(t)$ describing the amount absorbed with time can therefore be obtained by deconvoluting the clearance of the oral dose $o(t)$ with the clearance of the intravenous dose $i(t)$.

The state of absorption is characterised by the fractional absorption rate, k. At time t after dose administration, the total quantity absorbed is given by $\int_0^t a(t) \, dt$ and the amount remaining in the gut is

$1 - \int_0^t a(t)\, dt$. The absorption rate is therefore given by:

$$k = \frac{a(t)}{1 - \int_0^\infty a(t)\, dt} \tag{6.2}$$

which should remain appreciably constant with time. Measurement of the concentration of absorbed tracer in the blood is usually employed to measure the absorption of calcium and iron.

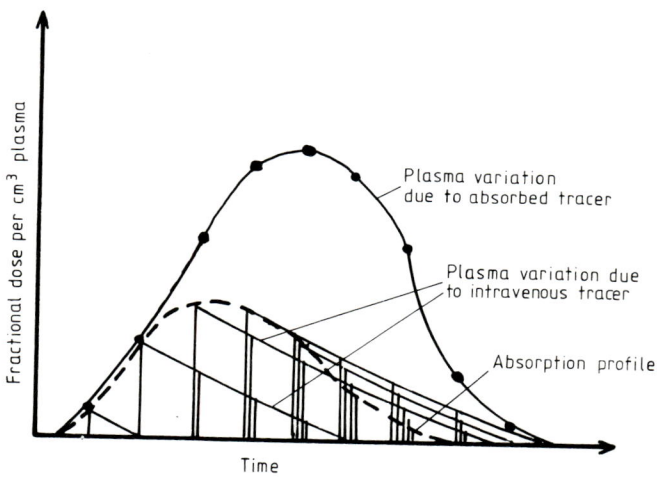

Figure 6.2   Relationship between the Ca absorption profile and the plasma clearance of absorbed tracer. The absorption of the tracer with time is indicated by the broken curve. Each point on this variation may be considered as the starting point of a plasma clearance curve whose shape is obtained from the clearance of an intravenous dose of Ca (full curves). The plasma variation for the absorbed Ca tracer (full circles) is obtained by summing the values of the overlapping clearance curves at each point.

The measurement of calcium absorption is important for establishing the cause of calcium deficiency and metabolic bone disease. Peak activity in the plasma can usually be detected about one hour after oral administration of a dose of 0.37 MBq (10 µCi) of calcium (^{47}Ca) chloride. Between 50 and 200 mg of stable calcium chloride is added to the dose as carrier. For the dual tracer technique described above, it is common practice to give 0.37 MBq (10 µCi) of calcium (^{45}Ca) chloride orally and 0.37 MBq (10 µCi) of calcium (^{47}Ca) chloride intravenously. Venous samples are withdrawn at intervals up to five hours with samples taken every 30 minutes in the first two hours and hourly thereafter. Since ^{47}Ca is a gamma emitting radionuclide and ^{45}Ca a beta emitting radionuclide, the clearance of the intravenous dose can be measured directly, by gamma counting using a well counter or automatic gamma counter (§3.4). Sufficient time is then allowed for the ^{47}Ca to decay (half-life 4.5d) before the ^{45}Ca content is assayed by liquid scintillation counting (§4.1). The drawbacks of this method are the need to use scintillation counting and the delay before results are available. An earlier indication of the state of absorption, though not an accurate result, can be obtained if the oral dose contains ^{47}Ca and the intravenous dose ^{45}Ca. In this way, the clearance of the oral tracer can be obtained promptly by gamma counting of the plasma samples, but this has not been common practice because the ^{47}Ca is normally supplied commercially in sterile form for intravenous use and the ^{45}Ca in non-sterile form. Reeve *et al* (1976) have avoided the need for liquid scintillation counting by using two gamma emitting radionuclides, ^{47}Ca for the intravenous dose and ^{85}Sr (as chloride) for the oral dose. These authors have shown that over the early period following dose administration, the physiological behaviour of ^{85}Sr is sufficiently close to that of ^{47}Ca for it to be substituted without substantial error.

The absorption of iron is sometimes measured to see if malabsorption is a contributory factor to anaemia. However, interpretation of results is often difficult because absorption is affected by the state of body iron stores. The simplest test is to administer 0.37 MBq (10 µCi) of ferrous (^{59}Fe) sulphate orally with 5 mg of ferrous sulphate as carrier. The ^{59}Fe absorbed into the blood is utilised during erythropoiesis for haemoglobin synthesis and is incorporated into new red cells in the marrow. The release of these labelled cells into the blood is substantially complete 10 days after dose administration when a venous sample is taken. This is haemolysed with saponin and counted in a well counter together with a diluted aliquot of the given dose. The patient's blood volume is also measured using ^{125}I–HSA (§5.2), or estimated from the patient's height and weight. The fractional dose absorbed, f, is then given by:

$$f = \frac{\text{circulating radioactivity in whole blood}}{\text{dose}}$$

$$= \frac{ksB}{kdV} \frac{r_b}{r_s}$$

where k is a factor relating activity to count rate, r_b is the net count rate per ml whole blood, B is the blood volume, r_s is the net count rate per ml of the diluted standard, V is the volume of dilution of the standard, s is the volume of the dose used for the standard and d is the volume of the dose administered.

This technique assumes that the same proportion of the absorbed tracer enters the red cells in all patients. This is not correct. In normal persons, about 80% of the tracer that enters the blood will be incorporated in circulating red cells 10 days after dose administration; in patients with iron deficiency 100% of the tracer may be taken up. In some haemolytic anaemias the proportion may be 60% or less as some of the absorbed tracer takes part in ineffective erythropoiesis and is returned from red cells in less than 10 days. To take account of individual patient variations, a dual tracer technique must be employed with an additional intravenous dose to indicate the proportion of the absorbed dose that becomes incorporated into circulating red cells. Generally a dose of 1.85 MBq (50 µCi) of ferrous (^{55}Fe) sulphate is given orally and 0.37 MBq (10 µCi) of ferrous (^{59}Fe) sulphate intravenously. At 10 days, a venous blood sample is taken and both radionuclides counted, together with diluted standards. ^{55}Fe is a low energy beta emitter and must be assayed by liquid scintillation counting. The presence of the gamma radiation from the ^{59}Fe and substantial quenching from the red colour of the sample makes this a difficult task to do accurately. The proportion of the injected dose which appears in circulating red cells may then be used to calculate the total absorbed dose. The fractional dose absorbed k, is then given by:

$$k = \frac{r_0}{D_0}\frac{D_i}{r_i}$$

i.e.

$$k = \frac{r_0}{r_i}\frac{D_i}{D_0} \tag{6.3}$$

where r_0 is the net ^{55}Fe count rate per ml blood, r_i is the net ^{59}Fe count rate per ml blood, D_0 is the total radioactivity in the ^{55}Fe dose and D_i is the total radioactivity in the ^{59}Fe dose.

6.2.3   Whole body retention

Measurement of the radioactivity retained by the patient using a whole body counter is clearly only possible with gamma emitting radionuclides. Measurement of the absorption of a tracer is straightforward and, at its simplest, is based upon two measurements of patient radioactivity. An initial measurement is made immediately following oral administration of the dose, to provide a measure of the whole dose under patient counting conditions. A second measurement is made some time later when the unabsorbed tracer has had sufficient time to be excreted in faeces and the body's radioactivity is solely due to the portion absorbed. Account must be taken of the radioactive decay of the tracer during the interval between the two measurements. This is usually done by counting a suitable container containing the same radionuclide at the time of each measurement. This is termed a phantom and should contain an indeterminate but sufficient activity to yield accurate counts. This method also compensates for any small differences in the spectrometer settings of the whole body counter on the two occasions. The net patient radioactivity B on the occasion of the second measurement, corrected for radioactive decay, will be given by:

$$B = B_2 \frac{S_1}{S_2}$$

where B_2 is the net patient count, S_1 is the net standard count at the time of the initial measurement, and S_2 is the net standard count at the time of the second measurement. The fraction of the dose absorbed k, is given by:

$$k = B_1/B = (B_1/B_2)(S_2/S_1) \tag{6.4}$$

where B_1 is the initial net patient count. To calculate B_1, the patient background count rate P_1 is measured by passing the patient through the whole body counter before the dose is administered. It is not possible to measure the patient background on the occasion of the second measurement since the patient contains radioactvity. The patient background count rate on the second occasion is calculated in relation to any change in the room background count rate; i.e. the patient background count rate for the second measurement, P_2, is given by:

$$P_2 = P_1(R_2/R_1)$$

where R_1 and R_2 are the room background count rates at the time of the first and second measurements. The patient background count rate is generally a little higher than the room background count rate since the patient scatters natural radiation into the counter which would not normally be detected.

Whole body retention has been used to measure the absorption of calcium, iron and vitamin B_{12}.

Whole body counting following an oral dose of calcium (^{47}Ca) chloride shows a rapid decline in body activity over the first two–four days due to the excretion of unabsorbed tracer. Thereafter the activity declines slowly, due mainly to the loss of absorbed tracer by the excretion of endogenous calcium into the gut. If measurements are made over a period of seven days, the latter part of the retention curve, if plotted on a log–linear graph, can be linearly extrapolated back to the ordinate axis (figure 6.3). The intercept on the axis provides an index of calcium absorption which correlates well with results of the faecal excretion method (see §6.2.4) (Deller *et al* 1965). A variant of this technique is to count only part of the body, usually the forearm. Clearly the behaviour of ^{47}Ca in the forearm must be representative of the whole body.

The absorption of iron is measured using expression (6.4), with an interval of 10 days between the two measurements. A dose of 10 µCi or less of ferrous (^{59}Fe) sulphate is sufficient.

The absorption of radioactively labelled vitamin B_{12} (cyanocobalamin) is a common test of the absorptive function of the small intestine.

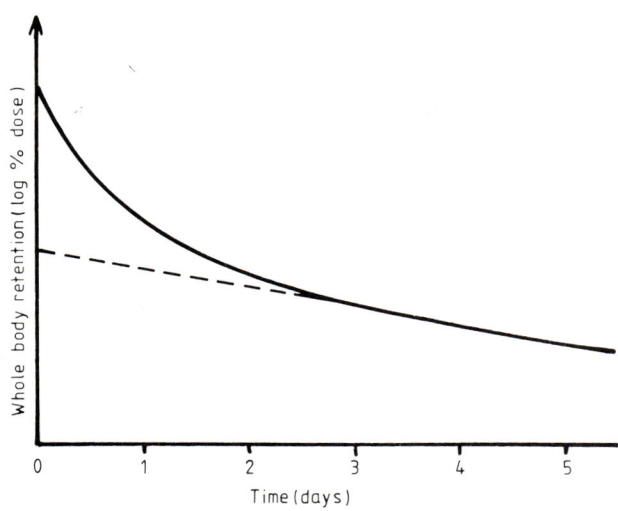

Figure 6.3   Whole body retention of an oral dose of ^{47}Ca. If the retention data are plotted on a log–linear graph, the latter part of the variation is linear and may be extrapolated back to the ordinate axis to give an index of calcium absorption.

Vitamin B_{12} requires gastric intrinsic factor (IF) for absorption and the B_{12}–IF complex is specifically absorbed in the ileum. Malabsorption may therefore be due to defective gastric function or to defective absorption. Malabsorption results in pernicious anaemia. This measurement is most commonly performed with the Schilling test (see §6.2.5). However, since this test requires a 26 hour urine collection which is usually obtained on an in-patient basis, measurement of the whole body retention was introduced to provide the test on an out-patient basis. A dose 37 kBq (1 μCi) of ^{58}Co cyanocobalamin is administered orally. Whole body counting takes places 30 minutes later and one week later, when the unabsorbed tracer has been excreted. The absorbed fraction is then calculated using expression (6.4). Normal subjects absorb more than 50% of the dose and patients with pernicious anaemia less than 20%. To determine if low absorption is a consequence of poor gastric function, the test may be repeated with the ^{58}Co cyanocobalamin bound to intrinsic factor. If a normal absorption results, the cause is poor gastric function, but if the absorption remains low the cause is poor ileum absorption.

In the author's experience, the results of whole body retention measurements do not always correlate well with the results of the Schilling test, which is well established. This discrepancy has been tentatively attributed to the possibility that retention may be altered by the state of the patient's store of vitamin B_{12} which is limited in its capacity. If the store is low the vitamin B_{12} absorbed will be retained and the retention will be indicative of the absorptive state. Conversely if the store is replete, excess vitamin B_{12} absorbed is excreted in urine and the amount remaining at seven days is less than that absorbed. This leads to an underestimate of the amount absorbed and to a misdiagnosis of malabsorption.

6.2.4 Faecal excretion method

This method requires the counting of daily faecal collections, typically over a period of seven days, to estimate the portion of the oral dose not absorbed by the body. The radioactivity excreted is then subtracted from the radioactivity in the dose to give the radioactivity absorbed. For accuracy the collections must be complete and this requires fastidious nursing care. The collections are counted on a large volume counter (§3.4.3) and may need to be homogenised to ensure good counting geometry. The standard containing a known aliquot of the dose should be made up in a container like that used for the samples and counted under the same conditions. The fraction of the dose absorbed, k, is then given by:

$$k = 1 - \frac{\Sigma_i C_i}{r_s \, d/s} \tag{6.5}$$

where C_i is the net count rate due to the ith faecal collection, r_s is the net count rate due to the standard, d is the volume of the dose administered, and s is the volume of the dose used in the standard.

This method of measuring absorption has been largely superseded by the measurement of whole body retention (§6.2.3) since the latter provides a direct measurement and avoids the errors arising from incomplete collections and the unpleasantness of handling faecal samples. It has been used in the past to measure both calcium and iron absorption. To measure calcium absorption, an oral dose of 0.37 MBq (10 µCi) of ^{47}Ca is given and faeces collected for four to six days. Correction has to be made for the endogenous secretion into the gut of absorbed ^{47}Ca; this amounts to three per cent of the dose. A correction for the endogenous secretion can be calculated specifically if a dual tracer technique with oral and intravenous doses is used. To measure iron absorption, a dose of 185 kBq (5 µCi) ferrous (^{59}Fe) sulphate is used and faecal collections are continued for seven to ten days.

6.2.5   Urinary excretion method

The most common example of using urinary excretion of an absorbed compound to indicate the amount absorbed is the Schilling test (Schilling 1953). As outlined in §6.2.3, this uses radioactively labelled vitamin B_{12} to measure gastric function and absorption at the ileum.

After an overnight fast and after emptying the bladder, the patient receives 18.5–37 kBq (0.5–1 µCi) of ^{57}Co or ^{58}Co cyanocobalamin orally. A 26 hour urine collection is also started. Two hours after receiving the dose, the patient receives 1 mg of non-radioactive vitamin B_{12} intramuscularly. This dose is sometimes called the 'flushing dose' since it flushes most of the absorbed vitamin B_{12} from the stores. This labelled vitamin B_{12} is excreted in the urine. When the urine collection is complete, an aliquot is taken for counting on a large volume counter. A known aliquot of the dose is also counted under the same conditions. The proportion of the dose excreted, k, is then given by:

$$k = \frac{r_u \, Vs}{r_s \, ud} \tag{6.6}$$

where r_u is the net count rate of the urine sample, u is the volume of the urine sample, V is the total volume of urine, r_s is the net count rate of the standard, s is the volume of the dose in the standard and d is the volume of the dose administered. Normal subjects excrete ten per cent or more of the dose and patients with pernicious anaemia five per cent or less. For patients in the second category the test should be repeated with a second dose of radioactively labelled cyanocobalamin bound to gastric intrinsic factor. If the proportion of the dose excreted then enters the normal range, the cause may be attributed to poor gastric function. If no change in the amount excreted occurs, the cause is malabsorption.

These two stages of the test have been combined into a single test using ^{58}Co cyanocobalamin and ^{57}Co cyanocobalamin bound to human gastric intrinsic factor (Bell *et al* 1965). The two doses are available commercially as capsules and are taken orally simultaneously. In the normal subject normal amounts of the two radionuclides are excreted. In pernicious anaemia, the patient excretes a reduced amount of ^{58}Co and more of the ^{57}Co. Patients with intestinal malabsorption excrete reduced amounts of both radionuclides. In counting the urine sample in this test, account must be taken of the contribution of ^{58}Co to the ^{57}Co counting channel. The Schilling test has been described in greater detail by Mollin and Waters (1968).

6.2.6 Breath exhalation method

Abnormal absorption of fats is a common feature of many gastro-intestinal disorders. Radionuclidic measures of fat absorption are considerably simpler to perform than the metabolic balance studies required with chemical estimations. Absorption is measured by the oral administration of a radioactively labelled fat. The fraction absorbed may be estimated either from the radioactivity recovered from faecal collections over three days or from the radioactivity appearing in the blood over the eight hours following dose administration. However, this test has not proved reliable in routine practice. It assumes that the faecal radioactivity is due only to unabsorbed fat and that radioactivity only leaves the gut as a result of fat absorption. Whilst the former assumption would appear to be correct, there is evidence to suggest that the latter is not. It is thought (Grenier *et al* 1965) that de-iodination of the fat takes place in the gastro-intestinal trace and the label is absorbed as iodide. As a result, the fraction absorbed is increased and the absorption overestimated.

The difficulty with the release of the label can be overcome by the use of an intrinsic label, usually carbon-14. This technique was introduced by Schwabe *et al* (1962), but has only recently become widely used. It requires the oral administration of 0.185 MBq (5 μCi) of ^{14}C long-chain fatty acid. A carrier comprising $1 \, \text{ml kg}^{-1}$ body weight of corn oil is also drunk. If absorbed, the fatty acid is catabolised and ^{14}CO$_2$ released into the blood stream. This is exhaled through the lungs along with endogenously produced carbon dioxide. The breath is sampled for its radioactive content, at hourly intervals for six hours after dose administration, by asking the patient to breath into a liquid scintillation counting vial containing hyamine hydroxide which absorbs carbon dioxide. An indicator added to the solution changes colour when one mmol of CO$_2$ has been absorbed. Breathing into the sample container is stopped at this point, which is usually after about a minute. Liquid scintillator is then added to the vial. The sample vials, together with a vial containing an aliquot of the dose in scintillator, are then counted on a liquid scintillation counter (see §4.1.6). After quench correction the results are expressed as percentage dose per mmol CO$_2$. These values are multiplied by the body weight in kg to take account of endogenous CO$_2$ production and are graphed against sample time to provide a ^{14}CO$_2$ output profile like that shown in figure 6.4. This profile

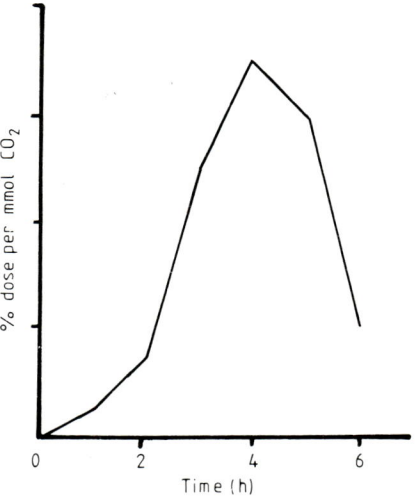

Figure 6.4 Exhalation profile for ^{14}Co$_2$ from the degradation of long-chain fatty acid. The outline is the five per cent profile for normal data. Profiles below the one shown are indicative of low absorption.

is based upon data from normal subjects and it shows the five percentile values at each time interval. Output profiles falling below this profile are indicative of poor absorption and profiles above it indicate normality. ^{14}C palmitic acid and ^{14}C oleic acid have both been used for this purpose.

The test has been extended by Kaihara and Wagner (1968) who used ^{14}C glyceryl tripalmitate instead of the ^{14}C fatty acid and showed a good inverse correlation between exhaled $^{14}CO_2$ activity and faecal fat. The tripalmitate is deconjugated by pancreatic enzymes in the gut to form palmitic acid and this is in turn absorbed and catabolysed. Exhalation of $^{14}CO_2$ therefore requires both the presence of pancreatic secretions and good absorptive function. The absence of the former gives rise to pancreatic steatorrhoea and the absence of the latter to intestinal steatorrhoea. Mills *et al* (1979) have routinely performed both tests with an interval of one week to provide an effective differential diagnosis between these conditions for screening purposes. The ^{14}C tripalmitate test is performed first; a low $^{14}CO_2$ output profile indicates the presence of a disorder. One week later the ^{14}C palmitic acid test is performed. A normal $^{14}CO_2$ profile indicates normal absorption and therefore a lack of pancreatic secretions. Another low $^{14}CO_2$ output indicates malabsorption and intestinal steatorrhoea. In this study, the maximum $^{14}CO_2$ levels in each profile were found to provide the greatest discrimination.

6.3   Clearance Measurements

6.3.1   Introduction

Measurement of the clearance of a tracer from the blood has already been mentioned in the context of other measurements. These were the correction for the extravascular diffusion of iodinated human serum albumin in the measurement of blood volume (§5.2) and the correction for the elimination of absorbed calcium in the measurement of calcium absorption from the gut (§6.2.2). However measurement of the clearance itself is diagnostic under certain conditions, which will be outlined later.

In measuring the blood clearance of tracers, the blood is generally assumed to be a simple dynamic system. The stable material being traced is assumed to leave the system at a constant rate and to be replaced by an in-flow of stable material at an equal rate. Thus the total amount of stable material remains constant and its movement can only be revealed by the addition of the tracer. If tracer is added to the system and equilibrates continuously throughout the system as it leaves with the out-flow, the amount of tracer leaving per unit time will be proportional to the amount remaining, i.e.

$$\frac{dA}{dt} = -kA \tag{6.7}$$

where dA/dt is the instantaneous loss rate of radioactivity, k is a constant and A is the radioactivity remaining. This is the standard expression for exponential loss of the tracer with time, whose solution is given by:

$$A = A_0 e^{-kt} \tag{6.8}$$

where A_0 is the initial activity at time zero. k is termed the 'fractional turnover rate' and gives the fraction of the system replaced per unit of time used for t.

The clearance of the tracer can be followed by plotting the activity of sequential blood samples as an exponential decrement with time on a linear plot like that shown in figure 6.5(*a*), or more commonly as a linear decrement on a log–linear plot like that shown in figure 6.5(*b*). The latter is preferred since it enables an accurate determination of the clearance from the best fitting straight line to the data. This is fitted using linear regression techniques.

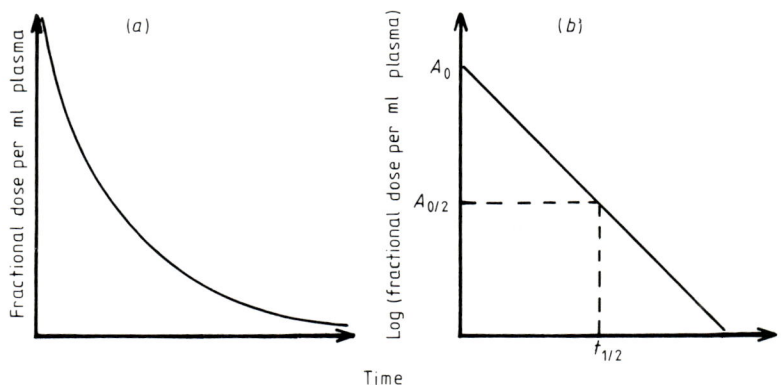

Figure 6.5 Expotential clearance of tracer (*a*) on a linear plot and (*b*) on a log–linear plot. The half-time is easily measured on the log–linear plot.

Instead of using the fractional turnover rate, k, clinical results are often expressed in terms of the half-life, or more properly, the half-clearance time $t_{1/2}$. The relationship between the half-time and k is readily determined by substituting $A = A_0/2$ into expression (6.8), i.e.

$$A_0/2 = A_0\, e^{-kt_{1/2}}$$

whence

$$t_{1/2} = (\ln 2)/k = 0.693/k. \tag{6.9}$$

The half-time may be determined readily from the log–linear plot (figure 6.5(*b*)) and used to calculate the turnover rate, k.

6.3.2 Plasma iron clearance and transport rate

The clearance of plasma iron can be measured by following the disappearance of an intravenous dose of ferric (^{59}Fe) citrate. A dose of 370 kBq (10 μCi) is used. In the plasma, the ferric iron is bound to transferrin, a plasma protein, and leaves the plasma for erythropoietic tissues where it is utilised for haemoglobin synthesis in red cell production. In patients whose transferrin is already saturated with iron, the tracer has to be incubated *in vitro* with unsaturated donor plasma prior to injection in order to obtain a sensible result.

Serial venous blood samples are withdrawn at 10 min intervals for 30 min after dose administration and thereafter at 30 min intervals to 3 h. After centrifuging, the supernatant plasma of each sample is counted in a well or automatic scintillation counter and the net count rate per ml plasma is plotted against sample time on a log–linear plot. Up to three hours, these points should lie on a straight line, although the last point may sometimes lie slightly above the line. The half-time may be determined directly from this graph. In normal subjects the half-time is 90 ± 30 min. Longer times are observed in patients with hypoplastic anaemia with low erythropoietic activity and red cell production and therefore a low utilisation of iron. Shorter times are observed in patients who are deficient in iron or have increased erythropoietic activity. The latter occurs in polycythaemia when excess red cells are produced and in haemolytic anaemia to replace prematurely sequestered or damaged red cells.

The half-time may be substituted in equation (6.9) to calculate the fractional turnover, k, i.e. the fraction of the plasma iron pool replaced per unit time. If the concentration of stable iron in the plasma is measured by chemical means, usually atomic absorption spectroscopy, the flow rate of iron through the plasma may be calculated. This is termed the plasma iron transport rate, F:

$$F = kM$$

where M is the total plasma iron. M is given by the product of the plasma volume V and the serum iron concentration s, i.e.

$$M = sV$$

and by substitution for M

$$F = ksV = 0.693\, sV/t_{1/2}. \tag{6.10}$$

Attempts have been made in the past to relate the plasma iron flow to the rate of red cell production and thence to red cell lifespan knowing the circulating red cell mass. (e.g. Silver 1962). This assumes that the plasma iron flow equals the flow into erythropoiesis and to a minor extent into long term iron stores. However it has been subsequently shown that the flow of iron through the plasma is the sum of three flows; into effective erythropoiesis and red cell production, into the lymphatic system from which it subsequently returns and into ineffective erythropoiesis from which it may return much later (Cook *et al* 1970). In normal subjects, the first of these flows is considerably greater than the other two and the plasma iron transport rate reflects the flow into erythropoiesis. However, in patients with sideroblastic and megaloblastic anaemia where there is substantial ineffective erythropoiesis, and plasma iron flow considerably overestimates the flow of iron into erythropoiesis.

6.3.3 Red cell survival

Measurement of the lifespan of red cells in the circulation is important for the diagnosis and assessing the severity of haemolytic anaemia. The common technique is to use red cells labelled with ^{51}Cr using the method of Gray and Sterling (1950) described in §5.3. Venous blood samples are taken starting on the day after injection of the labelled cells. These samples are haemolysed with saponin to ensure homogeneous distribution of radioactivity and known volumes of each are counted using a well or automatic scintillation counter. The net count rate for each sample is expressed as a percentage of the count rate of the sample taken on the day after dose administration, and the percentage values are plotted against time, usually on log–linear graph paper. Sampling usually takes place daily for three to four days, especially if severe haemolysis is expected, and thereafter every second day until the net count rate has fallen to 50% of the initial count rate. It is good practice to measure the haematocrit of each blood sample. Then in patients with falling haematocrits, the percentage values may be divided by the haematocrit to correct for the falling red cell mass in each sample.

The labelled cells will have ages ranging from virtually zero up to the full lifespan since they originate from a homogeneous venous sample. The concentration of the radioactivity will therefore decrease as the older cells are sequestered by the spleen. In the absence of pathological loss mechanisms, a constant fraction k of the cells will be lost per day. If N is the number of labelled cells in circulation, the loss rate will be given by:

$$\frac{dN}{dt} = -k.$$

If this expression is integrated between the limits $N = N_0$ when $t = 0$ and $N = 0$ when $t = T$, where N_0 is the initial number of cells and T is the lifespan, it can be shown that the disappearance is given by:

$$N = N_0\left(1 - \frac{t}{T}\right). \tag{6.11}$$

This is a linear expression relating N and t and if plotted on linear graph paper, the intercept on the abscissa can be used to determine the lifespan T, (figure 6.6) which ranges from 110 to 120 days.

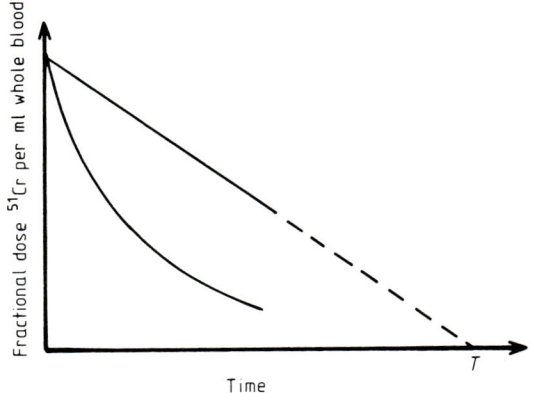

Figure 6.6 Survival of ^{51}Cr red cells. The linear relationship corresponds to a fractional loss rate and the curve corresponds to random losses. The line may be extrapolated to the abscissa where the intercept is the red cell lifespan T.

In the presence of pathological processes, the cells are destroyed at random either by external processes (extrinsic haemolysis) or by internal defects (intrinsic haemolysis). This affects labelled cells of all ages and the loss will be proportional to the number of cells present, i.e.

$$\frac{dN}{dt} = -kN.$$

Integration of this expression yields the familiar expression for exponential loss, i.e.

$$N = N_0 \, e^{-kt}. \tag{6.12}$$

This relationship is non-linear (figure 6.6) and the results must be plotted on log–linear paper in order to fit a straight line to the data. From this, the half-life can be obtained for clinical use.

In practice the survival curves do not reflect the disappearance of ^{51}Cr–red cells, but the disappearance of ^{51}Cr from the circulation. Unfortunately, the ^{51}Cr label is impermanent and elution from the red cells takes place. This takes two forms. During the first day after injection, a considerable loss of label (5–10%) takes place (early loss). This has been attributed to red cells damaged during the labelling process and promptly sequestered by the spleen. It is for this reason, that the 100% sample is taken one day after dose administration. There is also a steady loss of label which continues throughout the study. Elution makes it difficult to translate the data into a lifespan measurement. It is conventional practice to plot the percentage values against time on linear paper and observe the time at which the initial activity has fallen to one half. In normal subjects, the mean half-time is 28 days, with a range from 25–32 days.

The steady elution of the ^{51}Cr label has been measured by a number of workers and shown to be one to two per cent per day. For example, Bentley *et al* (1974) have compared the survival of ^{51}Cr labelled cells with DF^{32}P labelled cells in which the ^{32}P is fixed. From this comparative data, it is possible to prepare tables to correct the raw ^{51}Cr data and produce linear survival data in normal subjects, like that shown in figure 6.6, from which to determine the lifespan. However the validity of applying such a correction in pathological cases is dubious. Widman and Powsner (1974) developed a red cell model in which both the destruction rate and the elution rate were variable, in an attempt to fit individual ^{51}Cr clearance curves better. Whilst the data could be satisfactorily fitted in most cases, there were a number of cases in which a good fit to the data was impossible, suggesting that elution can be variable and unpredictable.

To overcome the inaccuracies and difficulties in determining red cell lifespan using ^{51}Cr, efforts have been made to measure the lifespan using ^{59}Fe. This has the advantage of being an intrinsic label, which is incorporated naturally into red cells. Direct measurement is clinically impracticable since the ^{59}Fe is incorporated into a cohort of new red cells (see §6.4.5) which may survive for up to 120 days. Moreover the label is recycled when the original cells die and it is difficult to observe the senescence of these cells. Alternative shorter methods of measuring red cell lifespan precisely using ^{59}Fe have been developed by Dagg *et al* (1972) and Ricketts *et al* (1975), but have not proved acceptable in routine use.

6.3.4 Renal clearances

Renal plasma clearance is defined as the volume of plasma completely cleared of tracer by the kidneys in unit time. As such, it is a useful rather than physiological concept. A greater volume of plasma may flow through the kidneys in unit time but can only be partially cleared of the tracer. Two measurements of total kidney performance are commonly made; (a) glomerular filtration rate (GFR) using a tracer which is removed solely by the glomerulus, and (b) effective renal plasma flow (ERPF) using a tracer which is cleared by glomerular filtration and tubular secretion.

The traditional method of measuring GFR uses inulin, a large polysaccharide, and this has become established as the reference technique. Inulin is infused intravenously at a constant rate and when a steady state has been reached measurements are made of the plasma concentration P and urine concentration U. The volume of urine passed in a given time is also measured and the urinary flow V calculated.

The output of the kidney O is then given by:

$$O = UV$$

which must be balanced by the input I given by:

$$I = CP$$

where C is the volume of plasma cleared per unit time. Combining these two expressions:

$$UV = CP \qquad \text{or} \qquad C = UV/P. \tag{6.13}$$

The concentrations of inulin in plasma and urine have to be measured chemically and this facility is not readily available. As a result tracer methods with their simpler analytical techniques have been developed.

[51]Cr–EDTA (ethylene diamine tetra-acetate) is the most common radio-active tracer for GFR measurements. This material is excreted almost exclusively by glomerular filtration, less than 0.5% being protein bound (Garnett *et al* 1967). During the test, a priming dose of 1.85 MBq (50 µCi) is given intravenously followed by the infusion of a solution containing 37 kBq ml^{-1} (1 µCi ml^{-1}) at a rate of 0.5 ml min^{-1}. After about 40 minutes, the plasma concentration becomes constant. A urine collection lasting about 15 minutes is then started and a venous sample taken at the mid-time. This process is repeated with rapid separation and counting of the plasma radioactivity until constant plasma activity is observed in two successive samples. The values of the urine and the plasma concentrations and the urine flow are then substituted into equation (6.13) to give the clearance. When the urinary flow is low, it may be necessary to catheterise the bladder in order to remove the whole of the urine sample for a particular time period.

Because of the complexities of the infusion technique, a single injection technique has been developed and evaluated. In the method of Chantler *et al* (1969), a single intravenous dose of 3.7 MBq (100 µCi) of ^{51}Cr–EDTA is given. Venous samples are taken at three, four, five and six hours after dose administration with another at 24 hours if renal failure is suspected. The venous samples are spun and the plasma separated and counted together with an aliquot of the given dose. The net plasma activities are then expressed in terms of fractional dose and plotted against time on a log–linear plot. A regression line is then fitted to the data and the line extrapolated back to the ordinate axis (see figure 6.3). The turnover rate k is determined from the slope of the line. The apparent distribution volume of the tracer V is obtained by dividing the count rate due to the administered dose by the plasma concentration given by the intercept on the ordinate axis (cf §5.2). The plasma clearance C is then given by:

$$C = kV. \qquad (6.14)$$

As a result of comparison of this method with standard inulin clearances, Chantler *et al* (1969) have modified expression (6.14) to:

$$C = 0.93 \, kV. \qquad (6.15)$$

The factor 0.93 takes account of three effects; the discrepancy between inulin and ^{51}Cr–EDTA clearances, the overestimate of the distribution volume by using a single exponential variation instead of the more complex true plasma clearance curve and the use of venous and not arterial blood samples to record the clearance.

For the measurement of effective renal plasma flow, para-aminohippuric acid (PAH) is the accepted material. 20% of PAH is eliminated by glomerular filtration and 80% by tubular secretion. It is generally thought that during one passage through the kidneys 90% of the PAH present is eliminated. To avoid cumbersome chemical estimations, a radioactive tracer is more commonly used. This is sodium (^{125}I) iodohippurate, which has a clearance 88% of that of PAH. Measurements of ERPF take the same form as those for GFR. Using the constant infusion method, a single dose of 740 kBq (20 µCi) is given followed by infusion of a solution containing 37 kBq ml^{-1} at a rate of 4 ml min^{-1}. For the single injection method, a dose of 1.48 MBq (40 µCi) is used.

Simultaneous measurements of GFR using ^{51}Cr–EDTA and ERPF using sodium (^{125}I) iodohippurate can be made. The contributions from the two nuclides in the plasma samples can be separated because the two nuclides have widely separated γ-ray energies.

6.4 Measurement of Organ Uptake

6.4.1 Introduction

The measurement of organ uptake of a radionuclidic tracer is the single most important measurement in nuclear medicine and therefore the most commonplace. A normal result is indicative of a patent blood supply and normal cellular function in the organ. An abnormal result is indicative of pathology. Because radionuclidic techniques render changes in function immediately apparent they can reveal pathological conditions long before morphological techniques. The latter reveal changes in tissue structure which may only become apparent after the pathology has been present for a considerable time. Coupled with imaging techniques, the uptake can be an index of regional organ function; this information is not provided by any other non-invasive technique. Serial uptake measurements with time, with or without images, considerably enhance the information content of the result and provide better diagnostic information.

Uptake measurements may be qualitative, as in most organ imaging, or quantitative. The latter category includes single measurements of organ uptake where the radionuclidic content is changing slowly and serial measurements where the radioactivity is passing through the organ. These different types of measurement are discussed separately below.

6.4.2 Organ imaging

Organ imaging is the most common investigation in nuclear medicine. About 85% of doses of radiopharmaceuticals are used for this purpose at the present time. The radiopharmaceuticals used for organ imaging are designed as far as possible to concentrate in the organ of interest. The design takes account of any special physiological or biochemical features of the organ. After intravenous injection of the radiopharmaceutical, time is allowed for it to concentrate in the organ. The distribution of radioactivity in the organ is then imaged with a rectilinear scanner (§3.7.1) or a gamma camera (§3.7.2). The relative uptake of the radionuclide in the different parts of the image and the appearance of the image relative to a normal image are used subjectively to assess organ function. This assessment can be made less subjective in suitable cases by image quantification (§3.7.3), e.g. by using regions of interest or count profiles across the organ.

About 90% of the doses of radiopharmaceuticals for organ imaging contain the short-lived radionuclide ^{99m}Tc, obtained from the ^{99}Mo–^{99m}Tc generator(§1.5). Common ^{99m}Tc labelled radiopharmaceuticals are given in table 6.1, together with dosage and usage. Other radiopharmaceuticals used for imaging are listed in table 6.2. The mechanisms of concentration in each organ are described briefly below.

Brain images are obtained with a radiopharmaceutical which is distributed in extracellular fluid, including the plasma, and therefore has a relatively low concentration in brain tissue. When an intracranial lesion is present, e.g. a tumour, infarct or abscess, there is a larger extracellular fluid space associated with the lesion and its surrounding. Also the blood volume of the lesion is greater than in normal tissue. As a result of these two factors, there is an increased uptake of tracer. An abnormal feature in the brain is therefore shown as a zone of increased activity over its surroundings. Sodium (^{99m}Tc) pertechnetate is commonly used, and to a lesser extent ^{99m}Tc–DTPA. The former has the slight disadvantage that perchlorate also has to be given to prevent uptake in the choroid plexus which might otherwise obscure pathological features. ^{99m}Tc–DTPA has the advantage that it is not concentrated by the choroid plexus but a larger dose has to be given to compensate for its more rapid excretion.

Liver and spleen imaging relies on the intravenous administration of small radioactive particles and their removal from the blood by phagocytosis in the Kupfer cells of the reticulo-endothelial system. The principal agents are tin and sulphur colloids with a particle size of

Table 6.1 Common [99mTc] labelled radiopharmaceuticals for organ imaging.

Radiopharmaceutical	Dose (MBq)	(mCi)	Use
sodium pertechnetate	37	1	Thyroid imaging
sodium pertechnetate	370	10	Brain imaging
DTPA[1]	74	2	Static renal imaging
DTPA[1]	185	5	Dynamic renal imaging
DTPA[1]	370	10	Brain imaging
DMSA[2]	74	2	Static renal imaging
HSA[3]	740	20	Cardiac ventricular function
MAA[4]	74	2	Lung perfusion imaging
MDP[5]	555	15	Skeletal imaging
Red cells	740	20	Cardiac ventricular function
Tin or sulphur colloid	74	2	Liver and spleen imaging
Tin or sulphur colloid	1.85	0.05	Lacrimal drainage

[1]Diethylene triamine penta-acetate.
[2]Dimercapto succinic acid.
[3]Human serum albumin.
[4]Macro-aggregated albumin.
[5]Methylene diphosphonate.

Table 6.2 Other radiopharmaceuticals for organ imaging.

Radiopharmaceutical	Dose (MBq)	(mCi)	Use
Gallium ([67Ga]) citrate	74	2	Localisation of neoplasms and infection
[75Se]–selenomethionine	7.4	0.2	Pancreas imaging
[75Se]–selenocholesterol	7.4	0.2	Adrenal imaging
[111In]–Calcium DPTA	37	1	Imaging of cerebro-spinal space
Sodium ([131I]) iodide	0.925	0.025	Thyroid imaging
Xenon ([133Xe]) gas	370	10	Lung ventilation imaging
Xenon ([133Xe]) gas in saline solution	370	10	Lung perfusion imaging
Thallium ([201Tl]) citrate	74	2	Myocardial imaging

400 nm. In diffuse liver disease, e.g. cirrhosis, there is diminished uptake throughout the liver and compensating increased uptake in other parts of the reticulo-endothelial system, e.g. the spleen or bone marrow. Space occupying lesions in the liver give rise to areas containing little radioactivity and often displacement of functioning tissue by comparison with the normal image.

Skeletal imaging relies on agents which are actively absorbed onto the hydroxyapatite crystals of bone. The degree of uptake is related to two factors, the rate of hydroxyapatite production (i.e. osteoblastic activity) and blood flow. Both these factors are elevated in a wide variety of bone pathologies and increased uptake occurs at the site of the lesion. However, because this happens in many pathologies, the test is not a discriminating one and cannot differentiate, for example, between metastatic and degenerative bone disease. Tracer which is not affixed to the skeleton provides a tissue background until excreted by the kidneys. This background reduces the visibility of the bone in the images. ^{99m}Tc–methylene diphosphonate is the current agent of choice since it has the most rapid blood clearance.

Renal imaging relies on tracers which are promptly removed from the blood by the kidneys. ^{99m}Tc–DTPA as a chelate is removed by glomerular filtration. It is used primarily for renal function measurements, and to a lesser extent as a scanning agent at early times after intravenous injection. ^{99m}Tc–DMSA is removed from the blood and fixed in distal tubules. Its distribution reflects the functioning kidney mass. Regions of the kidney with reduced function will contain less radioactivity.

Lung perfusion imaging uses radioactive particles with diameters in the range 20–50 μm which will not pass through capillary vessels. Following intravenous administration, they pass through the right heart and become trapped in the capillaries of the lung. The number of particles is limited so that typically only 1 in 10^4 capillaries is blocked. The most common agent is macro-aggregates of albumin labelled with ^{99m}Tc. These have a half-life of about seven hours in the capillaries so that the blockade is temporary. Lung perfusion imaging is used most often for the diagnosis of pulmonary emboli. The presence of an embolus in one of the arterial vessels in the lung prevents the particles entering that part of the lung and a non-radioactive region is apparent in the lung image.

Ventilation images of the lung are most often obtained using radioactive ^{133}Xe gas. An image showing the distribution of ventilation is obtained by the patient inhaling a bolus of the gas and then holding his breath whilst the image is obtained. If the patient then continues to breathe the gas on a closed respiratory circuit until an equilibrium concentration is achieved, an image in which the pattern of radioactivity is related to lung volume is obtained. If the patient then breathes in fresh air, the washout of xenon from the lung can be observed in serial images. These images reveal the ventilation rate of the lung, e.g. in poorly ventilated regions the concentration of radioactive gas will decrease relatively slowly. In some instances perfusion images of the lung are obtained using ^{133}Xe gas for direct comparison with the ^{133}Xe ventilation images. ^{133}Xe gas dissolved in saline solution is injected intravenously and when it reaches the lung, it rapidly enters the alveolar airsacs since the solubility of ^{133}Xe in air is 10 times greater than in blood. An image of the ^{133}Xe at this time reflects the distribution of perfusion.

For many years, imaging of the thyroid gland relied upon radioiodine which is concentrated by the gland because of its unique metabolic role in the formation of thyroid hormone. Sodium (^{131}I) iodine is the most suitable radiopharmaceutical containing iodine but the long half-life of eight days, and high radiation dose associated with ^{131}I, limits the activity that can be safely administered. Sodium (^{99m}Tc) pertechnate is also a suitable radiopharmaceutical for thyroid imaging. The pertechnetate ion TcO_4^- is concentrated by the thyroid gland since it has a similar electronic size to that of the iodide (I^-) ion. However it is not incorporated into thyroid hormone and images must be obtained at early times after dose administration (e.g. 20 minutes) before the tracer diffuses back into the blood stream. The larger activity of ^{99m}Tc that can be safely administered provides better quality images than those obtained with ^{131}I.

Images of blood volumes including measurement of cardiac function from serial ventricular images are obtained by using tracers which remain within the blood volume during the period of measurement. ^{99m}Tc labelled human serum albumin and red cells are the most common agents.

Imaging of the pancreas makes use of the synthesis of proteins from amino acids taking place inside it. A radioactive amino acid, ^{75}Se–selenomethionine in which ^{75}Se is substituted for a sulphur atom, is administered intravenously and takes part in the synthesis process. However, since protein synthesis is a distributed function, less than 10% of the dose concentrates in the pancreas. This low target specificity results in poor quality images.

Imaging of the myocardium can be done with potassium or its chemical analogues which are taken up by muscle mass. The most suitable radionuclide, on the basis of its radiation characteristics, is ^{201}Tl. Administration of the thallium (^{201}Tl) citrate takes place following exercise when any ischaemia will be present. The distribution of the ^{201}Tl then indicates the perfused zones of the myocardium. Late images are also obtained about six hours later when the ^{201}Tl will have diffused throughout the myocardial mass. Abnormal images then indicate the presence of lesions and the images used to qualify the results of the earlier exercise images.

Images of the adrenal glands may be obtained approximately seven days after the administration of ^{75}Se–selenocholesterol. It is common practice to obtain simultaneous images of the kidneys using ^{99m}Tc–DTPA to assist in locating the glands. This technique is useful for determining the presence of hyperactive glands.

Images showing the positions of infective pus and soft tissue neoplasms can sometimes be obtained from the concentration of gallium (^{67}Ga) citrate 48 hours after its administration. The mechanism of concentration for the ^{67}Ga is at present unknown.

The development of radiopharmaceuticals for organ imaging has been described in detail by Subramanian *et al* (1975). The clinical indications for organ imaging, imaging techniques and the appearance of images have been described by Maisey (1980) for the more common imaging investigations.

6.4.3   Static uptake measurements

These are quantitative measurements of the organ uptake of a radiopharmaceutical during which the uptake is constant at least for the period of the measurement. This is typically five minutes or less, for the convenience of the patient. Measurements are therefore confined to those parts of metabolic processes which proceed slowly. Uptakes can be measured either by organ imaging or by using a collimated scintillation detector connected to a scaler/timer.

When using organ imaging, image analysis with a region of interest facility is required (§ 3.7.3). This is most commonly employed for measurement of thyroid uptake of ^{99m}Tc or ^{131}I. The former being imaged at 20 min and the latter at 24 or 48 hours after dose administration. An image of the thyroid gland is obtained containing not less than 10^4 counts for one per cent accuracy of measurement. This is often not possible with a low uptake of ^{131}I and a time limit dependent on patient condition will need to be imposed. A rectangular region of interest is then outlined around the image of the gland and the count within the region determined. This count is the sum of tracer in the gland and tracer in plasma in the vasculature within the region of interest. The latter is often called the 'blood background'. To remove the contribution due to the blood background a second rectangular region of interest with about half the height of the gland region is set up below the gland and the count content determined. After normalisation to the area of the gland region, this count is subtracted from the gland count. A standard containing the same radioactivity as that administered to the patient is made up in the standard International Atomic Energy Agency (IAEA) phantom (IAEA 1962). This phantom is illustrated in figure 6.7. It is placed under the gamma camera with the simulated gland nearest to the collimator face. The separation between the phantom and the collimator face should be the same as that previously between the patient's neck and the collimator face. The standard is imaged for the same period of time as the patient and the net standard count is determined by subtracting the normalised count from an area adjacent to the image of the standard. The net patient count is then divided by the net standard count to provide the fractional uptake.

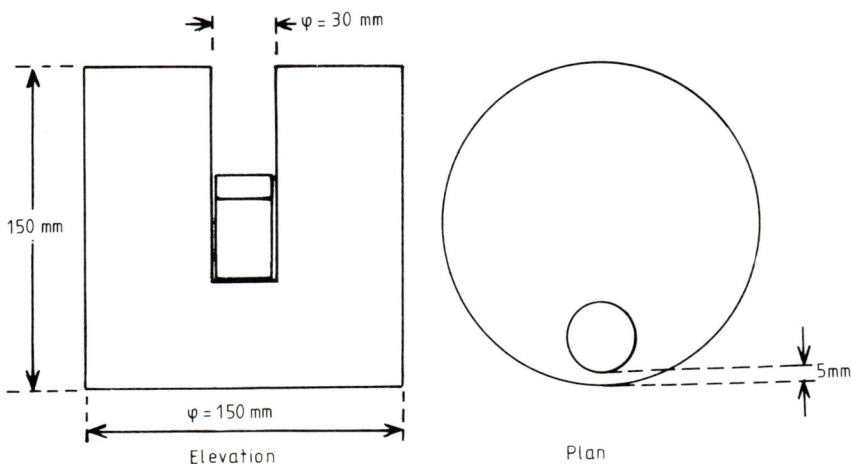

Figure 6.7   IAEA Neck Phantom.

For many years, thyroid gland uptake of ^{131}I has been measured using a collimated scintillation detector and scaler/timer. A diverging collimator like that shown in figure 3.10 is used. The collimator has been standardised (IAEA 1962) so that the field of view has a diameter of

120 mm at a distance of 250 mm from the face of the detector. The
scintillation detector should have a crystal not less than 25 mm diameter
× 25 mm thick. The patient is set at the pre-set distance from the
collimator and the count rate measured. The measurement is then
repeated with a thick lead shield interposed between the patient and
the detector to measure the background, including any contribution
from the patient that may penetrate the shielding laterally. This count
rate is subtracted from the first count rate to give the net thyroid count
rate. This procedure is then repeated using the IAEA standard (figure
6.7) containing the same activity as that given to the patient. The thyroid
uptake is calculated as described previously. At late times after dose
administration, which is the practice with ^{131}I, vascular activity is minimal
and no account of it need be taken. However at the earlier times
necessary when using sodium (^{132}I) iodide and ^{99m}Tc, account must be
taken of vascular radioactivity within the field of view of the scintillation
detector. A correction technique has been developed for these circumstances (IAEA 1972). The measurement of thyroid uptake in the diagnosis of thyroid disease is performed much less frequently now than in
the past. This is due to the successful development of radioimmunoassay
of thyroid hormones (Chapter 9). These techniques provide a direct
measure of thyroid hormone concentrations, are more convenient since
they are performed *in vitro* and do not present a radiation risk to the
patient.

Uptake measurements with a collimated scintillation counter are also
commonplace in haematology. A parallel hole collimator like that shown
in figure 3.10 is used. This technique is used for two purposes; (a) to
determine the sites of sequestration of ^{51}Cr–red cells and (b) to follow
the movement of ^{59}Fe in the circulation and during erythropoiesis. A
gamma camera is inappropriate in these cases because the low doses of
radioactivity which must be used provide insufficient photons for good
quality images.

The determination of the sequestration sites of ^{51}Cr–red cells is normally done alongside the determination of the red cell lifespan (§ 6.3.3).
Following administration of the ^{51}Cr–red cells (labelled by the technique
described in § 5.3), the collimated detector is sited over the heart, liver
and spleen on every second day. After subtraction of the background
count rate, in each case, the liver/heart and spleen/heart ratios are
calculated and plotted against time. The heart count provides a reference
to take account of radioactive decay and small differences in counting
conditions. In the absence of haemolysis, the liver activity should

decrease in parallel with the heart activity giving a constant ratio, whilst the spleen/heart ratio rises slightly due to the senescence of the older ^{51}Cr–red cells. In the presence of haemolysis, the spleen/heart ratio will rise steeply, and if the haemolysis is particularly severe the liver/heart uptake will also rise. Diagnostic criteria are normally based on the spleen/heart and liver/heart ratios at the ^{51}Cr half-clearance time.

Measurements of localised uptake of ^{59}Fe usually accompany the measurement of the plasma iron transport rate (§ 6.3.2). Measurements of the heart, liver and spleen and sacral activities are made at frequent intervals on the first day, daily for one week and every second day thereafter. The sacral count is a measure of bone marrow activity. This site is chosen because it is the only accessible one clear of major blood vessels which would otherwise make an appreciable, vascular contribution to the count rate. A 'standard' is also counted with each set of measurements to take account of radioactive decay and small changes in spectrometer settings. This is a container with a given activity (typically 37 kBq (10 µCi) of ^{59}Fe in solution and having a cross-section greater in area than that of the collimator. Each count is referred back to the first day using the expression:

$$C_n^1 = C_n \frac{S_0}{S_n}$$

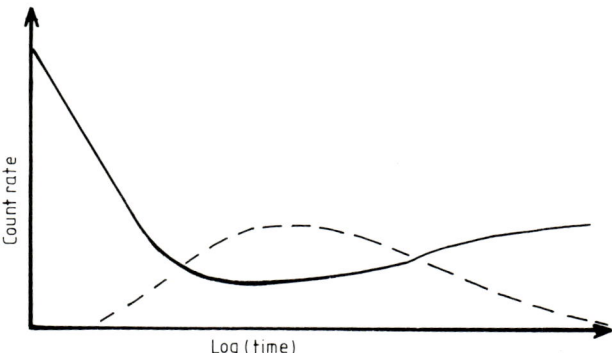

Figure 6.8 External counting of ^{59}Fe activity. The full curve is the variation in blood ^{59}Fe activity measured precordially. The clearance of the tracer from the plasma is first seen, followed eventually by its reappearance in labelled red cells. The complementary rise and fall in marrow ^{59}Fe activity measured over the sacrum as the ^{59}Fe takes part in erythropoiesis is also shown.

where C_n^1 is the corrected net count rate from day n, C_n is the measured net count rate on day n, S_0 is the net standard count rate on day 0 and S_n is the net standard count rate on day n. The values of C_n^1 for each site are plotted against time on a linear–log plot. Normal results are illustrated in figure 6.8, showing the initial disappearance of the tracer from the blood (indicated by the heart counts) into the marrow (sacral counts) and then its return to the blood. In normal persons, the spleen and liver counts are due solely to vascular activity and mimic the heart counts. In the presence of haemolysis, the spleen counts will rise at a late stage due to the sequestration of ^{59}Fe–red cells. If haemolysis is especially severe, an upturn in the liver counts may also be observed.

6.4.4   Dynamic uptake measurements

These are quantitative measurements of organ uptake in which the uptake varies continuously. The uptakes can be measured either by sequential organ imaging or by using a collimated scintillation detector connected to a ratemeter.

When using organ imaging, image analysis with a region of interest facility is required (§ 3.7.3). Each region of interest is selected on a good quality image obtained from summing a number of pertinent sequential images and is then applied to all images in the study. The variation in counts with time in the region of interest throughout the study are then displayed. Diagnosis is based upon the shape of the curve obtained and in some instances from fractional turnover rates derived from the slope of the curve. Common dynamic studies are considered briefly below.

In conjunction with static brain imaging, it is possible to observe the arrival of the tracer dose in the brain. The activity can be observed entering the carotid arteries and then filling the cerebral vasculature. It is easy to detect a delay in the filling on one side relative to the other, which may be indicative of arterial stenosis or a cerebro-vascular accident.

Dynamic renal imaging may be performed with ^{99m}Tc–DTPA or sodium (^{123}I) iodohippurate if the latter nuclide is available. The former is filtered by the glomerulus and the latter by both glomerular filtration and tubular secretion. The kidney activity curve peaks and then falls as the tracer is first concentrated from the blood and then excreted into the urine. The test lasts about 20 minutes. The phases of this curve have been examined extensively and comprise (a) an initial rapid rise when the vasculature of the kidney fills with radioactivity, (b) a so-called 'parenchymal' phase as the kidney concentrates the activity and (c) a washout phase as the kidney excretes the activity. In reality this is an oversimplification since the last two processes happen simultaneously and the peak represents the stage where the radioactivity entering equals that leaving. However, examination of the separate phases is helpful for diagnosis. A reduced height is indicative of poor renal function, a delayed peak time may be indicative of renal artery stenosis, and a slower excretion or even 'hold-up' of the tracer in the kidney may indicate an obstructive uropathy. The different activity–time variations obtained in these situations are illustrated in figure 6.9. When renal function is poor, account must be taken of the contribution of vascular activity to the renal uptake. This is normally corrected by using a background region of interest as described in § 6.4.3 for the thyroid gland. However, the best area to select for this purpose is not clear.

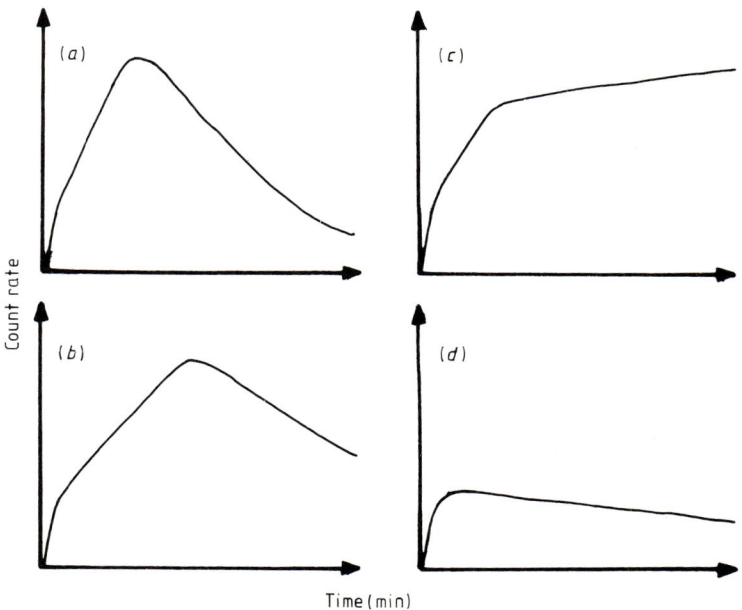

Figure 6.9 Radioisotope renograms: (*a*) shows normal function, (*b*) shows delayed peak activity due to renal artery stenosis, (*c*) denotes an obstructive uropathy which prevents the excretion of the tracer and (*d*) shows renal failure where function is poor and the curve largely reflects kidney vascular activity.

Dynamic renal imaging has also become important for the prompt and non-invasive assessment of renal transplants. A short one minute study using ^{99m}Tc–pertechnetate or the longer study described above using ^{99m}Tc–DTPA is performed. The short study is a test of the renal blood supply. In the functioning kidney, a sharp peak in kidney activity is observed as the injected bolus passes through. In the presence of immunological rejection or acute tubular necrosis, a slowly increasing retention of the tracer is observed.

Dynamic imaging of the liver is a relatively new technique using ^{99m}Tc labelled imido-acetic acid derivatives. These are concentrated by the polygonal cells of the liver and secreted into the gall bladder. The contents of the gall bladder may then be discharged through the common bile duct into the gastro-intestinal tract. These studies have a developing role in the investigation of the biliary tree.

With miniaturisation, image analysis computers have become more powerful and more complex techniques can be used. One of these techniques is multiple gated acquisition, which has made the accurate measurement of the left ventricular function of the heart possible. The patient's electrocardiogram (ECG) signal is fed into the computer which divides the interval between successive R waves into equal segments, most often 16. Heart images are then acquired for a given time or until a preset information density is reached. During this process, the images corresponding to each segment of the heart cycle are added to the previous images derived from that same segment. In this way good quality images are built up for each segment of the heart cycle and these may be analysed quantitatively. The commonest technique is to investigate the left ventricular activity during the heart cycle. This clearly represents the ventricular volume and is a minimum at systole and a maximum at diastole. The difference between maximum and minimum divided by the maximum is the ejection fraction; this is illustrated in figure 6.10. This technique has also been used to look for akinetic segments of the heart muscle, which may be the result of ischaemia.

Before the advent of the gamma camera, renal function was regularly assessed using two collimated detectors, each connected to a ratemeter. The collimators were of the parallel hole type and were sited one over each of the patients's kidneys whilst the patient lay face downward. The activity of each kidney was recorded for 20 minutes on a chart recorder connected to the outputs of the ratemeters following the intravenous injection of a dose of sodium (^{131}I) iodohippurate. This procedure was called the 'radionuclide renogram' and the kidney activity curves obtained are similar to those described above with the gamma camera

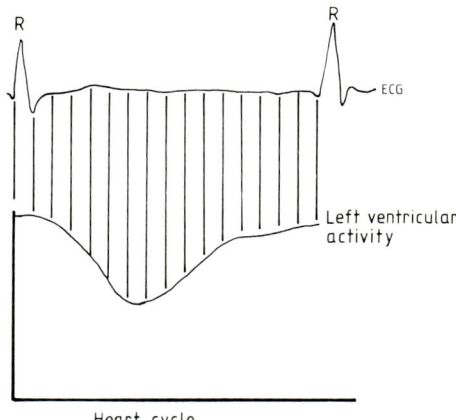

Figure 6.10 Multiple gated acquisition of cardiac images. The RR interval is divided into sixteen segments and the images obtained over successive heart cycles are superimposed to improve the statistical accuracy. The activity in the left ventricle may be measured over the heart cycle and the ejection fraction calculated from (maximum activity–minimum activity)/maximum activity.

(figure 6.9). Before performing the test, care had to be taken to equalise the sensitivities of the two detectors; this was usually done with small alterations to the window of the pulse height analysers. CABBS renography (Britton and Brown 1971) was employed to subtract the vascular contribution to the kidney activity in patients with suspected poor renal function. This technique employed a third collimated detector sited over the heart to follow the vascular activity. During the study, a fixed fraction of the instantaneous heart activity was continuously subtracted by analogue circuitry from each kidney's activity to give the true variations in renal activity. Before the study a 185 kBq (5 µCi) dose of ^{125}I–HSA was given intravenously and allowed to equilibrate throughout the patient's blood volume. The patient was then aligned beneath the three detectors and the subtraction fraction set for each kidney so that the heart count annulled the kidney counts in each case. This procedure empirically takes account of the different blood volumes in the fields of view of the detectors.

6.4.5 Physiological uptake measurements
There are two common uptake measurements which involve physiological entities rather than the anatomical entities studied so far. These entities are red cells and deep vein thrombi.

The red cell utilisation of ^{59}Fe is often performed in conjunction with the measurement of ^{59}Fe plasma clearance and the external counting of ^{59}Fe uptake (§ 6.4.3). At 7, 10 and 14 days after the administration of 370 kBq (10 µCi) of ferric citrate, venous blood samples are taken. Each sample is haemolysed with saponin and a known volume of whole blood counted together with a diluted aliquot of the given dose. The patient's blood volume is estimated from height and weight tables or by direct determination using ^{125}I–HSA and the haematocrit (§ 5.2). The ^{59}Fe red cell utilisation U on each day is then given by the activity of the tracer circulating in the red cells divided by the dose activity, i.e.

$$U = \frac{r_c B v S}{r_s V d}$$

where r_c is the count rate per ml whole blood, r_s is the count rate due to the aliquot of the standard, B is the blood volume (ml), v is the volume of the diluted standard counted, V is the dilution volume of the standard, S is the volume of the dose taken for the standard and d is the volume of the dose. In the normal person, the utilisation will rise from day 7 to day 10 and only a little thereafter. The maximal value will typically be about 80%. In the iron deficient patient, the utilisation rapidly rises to 100%. In patients with aplastic anaemia and little marrow erythropoietic activity, the utilisation may be 5% or less. In patients with haemolytic anaemia, utilisations around 60% are commonplace, because although utilisation is normal, some of the labelled cells have been removed from the circulation by the haemolytic process.

The formation of thrombi in the venous system involves the deposition of fibrinogen from the plasma. If ^{125}I–fibrinogen is administered intravenously at the time of thrombus formation, it too will be incorporated into the thrombus. The local concentration of ^{125}I activity can then be detected by a collimated scintillation detector held external to the body. In practice the low energy gamma radiation (30 keV) from ^{125}I limits this technique to the detection of thrombi in the legs. Measurements are made along the length of each leg using a small hand-held sodium iodide scintillation detector, daily for five days following administration of 37 MBq (100 µCi) of ^{125}I–fibrinogen. The measurements are usually made at regular intervals at sites which are marked on the leg and a 20% change in count rate between adjacent sites is regarded as diagnostically significant. This technique is useful in the assessment of post-operative patients in whom there is a risk of thrombus formation. A typical commercial detector and ratemeter are illustrated in figure 6.11.

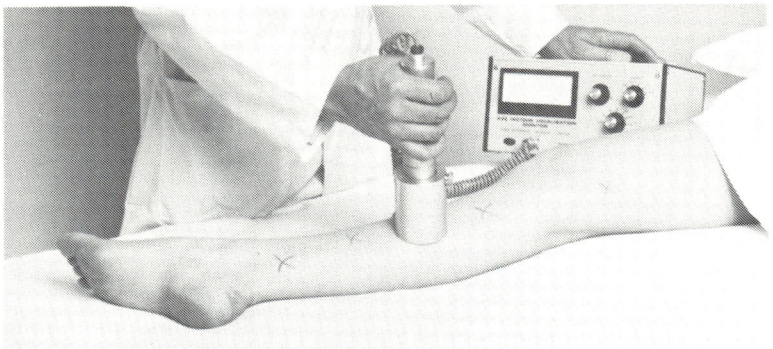

Figure 6.11 Deep vein thrombosis monitor. The lightweight scintillation detector is shown (centre) being held over the calf of the leg. The ratemeter can be seen (right). (Reproduced by kind permission of D A Pitman Ltd.)

6.5 Determination of Body Losses

Radioactive tracers are frequently used for two determinations of the body loss of materials. These are the loss of blood and the loss of protein from the gastro-intestinal tract.

To measure blood loss, a sample of the patient's blood is taken and labelled with ^{51}Cr, as described in § 5.3. The labelled cells are then re-injected into the patient and daily faecal collections started for a period of seven days. The collections are preferably taken straight into containers which can be counted on a large volume scintillation counter (§ 3.4.3). A daily venous blood sample is also taken and diluted to the same volume as the faecal samples. These are counted in the same counter to provide the count rate per ml whole blood on each day. If these values are divided into the corresponding faecal count rates, the blood loss on each day in ml can be calculated. Daily losses in excess of 2 ml are considered abnormal.

Protein loss from the gastro-intestinal tract has been measured with the greatest success when using ^{51}Cr–human serum albumin (^{51}Cr–HSA). This is administered by intravenous injection and daily faecal collections are made over the following five days. In the normal person, less than one per cent of ^{51}Cr is excreted in the faeces. The ^{51}Cr–HSA is unstable and should be prepared fresh prior to use. A good correlation has been observed between the faecal ^{51}Cr output and the severity of hypoalbuminaemia.

7 Measurement of Blood Flow

7.1 Introduction

Measurement of blood flow may be divided into four categories on an anatomical basis. These are:

(a) cardiac output,
(b) flow in major vessels,
(c) organ blood flow, and
(d) localised tissue blood flow.

Radionuclidic techniques have been used with varying degrees of success in these different categories. Accurate and reliable measures of blood flow all require a high concentration in the target vessel by comparison with the adjacent tissues at some point in the measurement. This occurs during the measurement of cardiac output since the whole of the injected dose passes through the chambers of the heart before entering the vessels and tissues. It also occurs during the measurement of local tissue blood flow if a freely diffusible tracer is used which is completely removed from the measurement site by the blood flow.

A poor differential between the activities in the target vessel and the surrounding tissue is often observed when detecting the radioactivity in major vessels. The peak vessel activity will always be lower than the peak cardiac activity since the latter will be distributed amongst a number of vessels and the bolus of activity will have been smoothed by the longer passage from the site of administration. The field of view of the collimated detector may also include other vessels. The background due to activity in the surrounding vasculature may be appreciable and always builds up rapidly. This is especially true when measuring low blood flow since collateral circulation may have developed. For these reasons, radionuclidic measurements of vessel blood flow have not entered routine practice. Techniques such as Doppler ultrasound show much greater promise.

The above difficulties, i.e. a low delivered activity and an appreciable vascular background, are also present in most attempts to measure organ flow. As a result, organ blood flow is not routinely measured using radionuclides. An exception where these drawbacks are less problematical is the brain, and cerebral blood flow has been extensively studied using radionuclides. This is because the major vessels to the brain—the carotid arteries—are accessible for dose administration and the tissue vasculature is limited to the cranium.

The measurement of cardiac output, cerebral blood flow and tissue blood flow are considered separately in the following sections.

Measurements are commonly made with a collimated scintillation counter connected to a ratemeter or a number of such systems. Single parallel hole collimators are normally used, the size of the hole being related to the size of the region of interest. Analogue ratemeters are used for graphical record of the passage of the tracer to provide a prompt indication of successful technique. A digital ratemeter or recycling scaler is often connected in parallel to provide a quantitative output for the determination of the slope or of the area beneath curves of tracer activity.

The tracer is used as a flow marker. Its chemical nature is therefore of secondary importance. Its physical nature is selected according to the experimental design of the measurement. This becomes clearer in the following examples.

7.2 Cardiac Output

Cardiac output is the volume of blood ejected by the heart per unit time. For haemodynamic equilibrium the volume ejected by the left ventricle must equal that ejected by the right ventricle, but measurements are usually confined to the latter.

The scintillation detector is sited precordially over the heart. A bolus of activity is injected intravenously into an arm vein and its passage through the left ventricle is recorded. The cardiac output is calculated from the area beneath the activity–time curve. This is illustrated in figure 7.1 where a collimated scintillation detector monitors a single blood vessel. If a single radioactive particle passes along the vessel, the activity–time curve has a 'top hat' shape like that shown in figure 7.1(*a*). If the flow rate increases, the particle passes more rapidly through the field of view of the detector and the interval T decreases. The height of the activity–time curve will be related to the strength of the radioactive particle. In practice, the tracer is not present at a single point but has been mixed and diluted to give a more diffuse zone of activity. However, each radioactive nucleus in the bolus can be considered to give an activity–time curve like that shown in figure 7.1(*a*) and these may be summed by the detector to give a response like that shown in figure 7.1(*b*). This assumes that the detector is equally sensitive to particles at all points in the cross-section of the vessel.

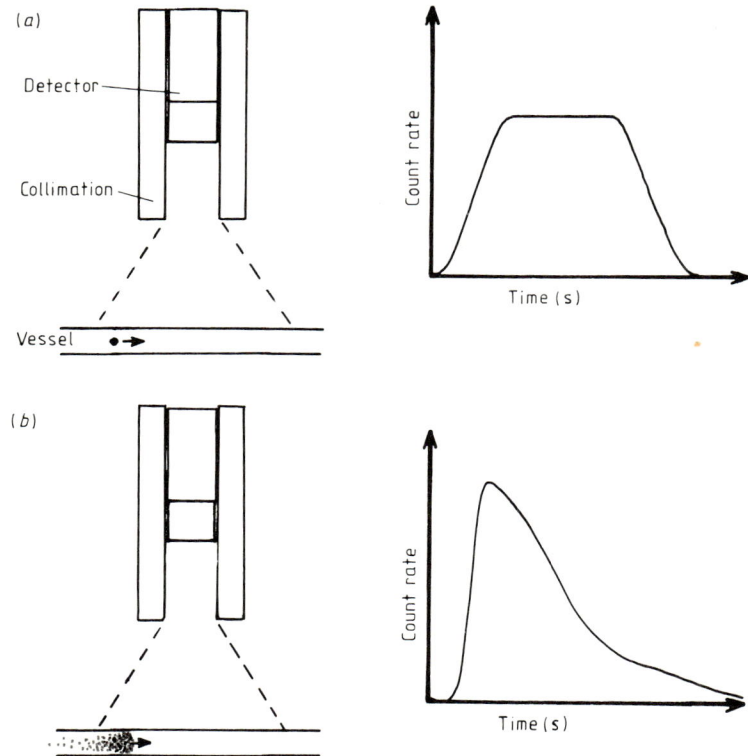

Figure 7.1   Counting Responses (*a*) to a single radioactive particle and (*b*) to a bolus of radioactive particles.

If one also assumes that the tracer is thoroughly mixed with the blood flow, the area beneath the activity–time curve is given by:

$$\text{area} = \frac{\text{dose}}{\text{volume flow rate}}.$$

If the ordinate is expressed in terms of fractional dose per litre and the activity–time curve is described by the function $c(t)$, this expression becomes:

$$\text{area} = \int_0^\infty c(t)\,dt = \frac{1}{\text{flow rate}}$$

i.e.

$$\text{flow rate} = \left(\int_0^\infty c(t)\,dt \right)^{-1}. \tag{7.1}$$

This is the Stewart–Hamilton dilution formula applied in a wide variety of measurements of cardiac output (Kinsman *et al* 1929).

In practice, the recorded activity–time curve does not look like that in figure 7.1(*b*) due to the effect of recirculation of the tracer after the first passage. A typical experimental curve is shown as a log–linear plot in figure 7.2. Since the left ventricle behaves as a well mixed system with a constant rate of out-flow, the contents will follow an exponential decline, as described in equation (6.8), until the appearance of the recirculated tracer. The contribution due to recirculated tracer can be excluded by extrapolating the initial linear decline to the abscissa. The area beneath the curve corresponding to the first passage of the bolus can then be calculated either graphically or numerically.

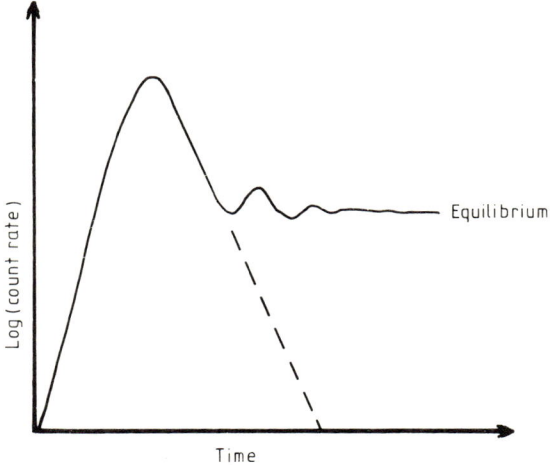

Figure 7.2   Cardiac output curve.

It is also necessary to calibrate the ordinate axis in order to express the area in units of fractional dose per litre per unit time. This is done by waiting until the tracer has equilibrated throughout the blood volume. The net precordial count rate, P, at this time, typically 10 minutes after dose administration, is noted and a venous blood sample is taken. This is counted in a well scintillation counter together with a diluted aliquot of the dose. The equilibrium blood activity is then expressed in terms of the fractional dose per litre whole blood, f, i.e.

$$f = \frac{C_s v a}{C_a V d s} \times 10^3$$

where C_s is the net count rate due to the blood sample, C_a is the net count rate due to the diluted aliquot, s (ml) is the volume of blood sample counted, v (ml) is the volume of diluted aliquot counted, V (ml) is the volume of dilution of the aliquot, a (ml) is the volume of the aliquot of the dose, and d (ml) is the volume of dose injected. The ordinate scale of the activity–time curve may then be converted to units of fractional dose per litre using the scaling factor f/p.

This method requires a non-diffusible tracer for two reasons. Firstly, implicit in equation (7.1) is the assumption that the whole of the administered dose passes through the heart. Secondly, the external count rate can only be related to blood tracer concentration if the tracer does not leave the circulation. A gamma emitting radionuclide must also be employed to permit precordial counting. Initially ^{131}I–HSA was used, but latterly this has been replaced by ^{99m}Tc–HSA or ^{99m}Tc–red cells and to a lesser extent by ^{113m}In. The last radionuclide labels the plasma protein transferrin *in vivo* if injected at an acidic pH.

Measurements of cardiac output can also be performed using a gamma camera with data analysis facility. This has the advantage that the left ventricle can be directly outlined and possible errors in positioning with a precordial counter are avoided. However, the camera must possess a linear response to count rate extending to high count rates in order to accurately record the first passage of the bolus through the heart.

7.3   Cerebral Blood Flow

Early techniques of measuring cerebral blood flow rely on application of the Fick principle (Klein 1930). This principle relates the amount of material removed from the blood stream by an organ to the concentrations of the material in the arterial and venous blood. The rate of uptake,

dQ/dt, is given by:

$$dQ/dt = F(C_a - C_v) \qquad (7.2.)$$

where F is the flow rate and C_a and C_v are the arterial and venous concentrations respectively. Integrating and rearranging, this expression becomes:

$$F = Q \left(\int_0^t (C_a - C_v) \, dt \right)^{-1}. \qquad (7.3)$$

It is not possible to measure Q directly. Kety and Schmidt (1948) overcame this difficulty by adopting an inert gas tracer. These tracers are freely diffusible due to their lypophilic properties and equilibrate virtually instantaneously with the tissue surrounding the blood vessel. The clearance of the tracer is, therefore, limited only by the blood flow. As a result the venous concentration $C_v(t)$ reflects the mean cerebral concentration C_b and is related to it by the blood to brain partition coefficient, λ, which has a value around 1.1, i.e.

$$\lambda = C_b/C_v. \qquad (7.4)$$

Equation (7.3) then becomes for 100 g of tissue:

$$F = 100 \, \lambda C_v(t) \left(\int_0^t (C_a - C_v) \, dt \right)^{-1}. \qquad (7.5)$$

By convention, blood flow is quoted for 100 g of tissue. Kety and Schmidt introduced this technique using 15% nitrous oxide, but it was later modified by Lassen and Munck (1955) using radioactive [85]Kr gas. The arterial and venous blood samples were drawn into planar cuvettes which were placed in contact with an end window Geiger counter for assay of their activity. Typical arterial and venous activity–time curves obtained by sampling at one minute intervals over a period of 15 minutes during inspiration of the gas are shown in figure 7.3. The values of $C_v(t)$ and the area between the arterial and venous curves may then be entered into expression (7.5). In normal subjects, the blood flow per 100 g tissue is about 50 ml min^{-1} corresponding to a cerebral blood flow of 700 ml min^{-1} for a 1400 g brain.

Clearly the method of using arterial and venous sampling with [85]Kr is very invasive and can only accompany carotid catheterisation. Less traumatic techniques using the gamma emitting radionuclide ^{133}Xe are now employed. This permits the non-invasive external detection of

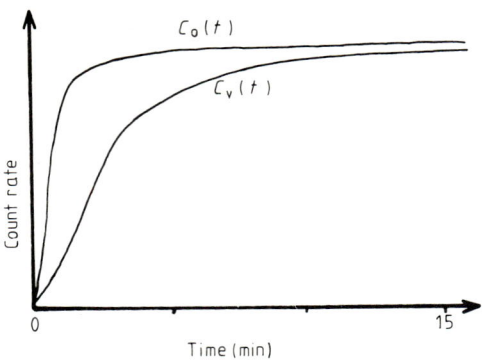

Figure 7.3   Arterial and venous concentrations of an inert gas tracer
during constant infusion of an organ.

cerebral radioactivity and also the measurement of regional cerebral
uptake using a number of collimated detectors. The ^{133}Xe may be
administered dissolved in saline for injection or by inhalation as a gas.

Using the injection technique the dose of 18.5 MBq (500 µCi) of ^{133}Xe
in 2 ml of saline solution is administered by carotid puncture. In general
practice, the activities in the cerebral hemispheres are detected with a
pair of collimated scintillation detectors and the activity–time variations
are recorded using ratemeter systems (figure 7.4). Recording continues
for about 15 minutes after dose administration. After one passage
through the brain, the tracer reaches the lungs via the left heart. Due

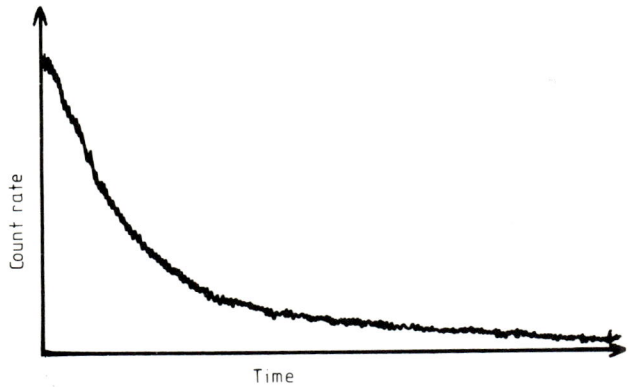

Figure 7.4   Clearance of an inert gas tracer from a cerebral
hemisphere.

to its ten-fold greater solubility in air than in blood, most of the ^{133}Xe is exhaled and a negligible amount returns to the brain in the arterial supply. If we consider equation (7.5) in the situation following the equilibration of the arterial bolus, then $C_a = 0$ and equation (7.5) becomes:

$$F = 100 \, \lambda C_v(t) \left(\int_0^t C_v \, dt \right)^{-1}. \tag{7.6}$$

$C_v(t)$ is the height of the activity–time curve and $\int_0^t C_v \, dt$ is the area beneath it. This leads to a very ready method of calculating the mean cerebral blood flow if these values are substituted in equation (7.6). This is often termed Zierler's height over area method and is due to Zierler (1965).

Alternatively the brain can be considered as two compartments comprising white and grey matter respectively. Each compartment will show the exponential decrease described by equation (6.8). It is assumed that the late part of the cerebral clearance curve is due to the slower component and this is extrapolated linearly back to the ordinate axis on a

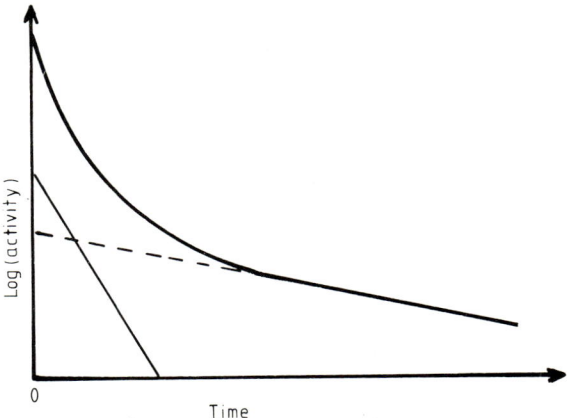

Figure 7.5 Analysis of tracer clearance using curve stripping. On a log–linear plot, the late part of the curve is assumed to be due only to the slowest exponential variation, the contributions from the faster exponentials being assumed to be negligible. This late part is extrapolated back to the ordinate axis (broken line) and the differences between it and the observed variation are plotted (full line). In this example, it yields a single exponential, but this could be a multi-exponential variation that would require stripping in a similar way.

log–linear plot (figure 7.5). The differences at intervals between the extrapolated line and the early part of the clearance curve are calculated and plotted. These too should lie on a straight line. The fractional rate constants K (equal to $0.693/t_{1/2}$) for each line are calculated. This process is known as 'curve stripping' and may be performed for a number of superimposed exponential variations.

The clearance of the tracer from the tissue can be calculated very simply from the fractional clearance rate k. Provided diffusion equilibrium occurs, the rate of decrease in activity in 100 g of tissue is given by:

$$100 \frac{\mathrm{d}C}{\mathrm{d}t} = F C_v \tag{7.7}$$

where C and C_v are the tissue and venous concentrations of tracer respectively and F is the blood flow. However, from equation (7.4):

$$C_v = C/\lambda.$$

Also

$$\frac{\mathrm{d}C}{\mathrm{d}t} = kC.$$

Eliminating $\mathrm{d}C/\mathrm{d}t$ and C_v from equation (7.7), one obtains

$$- 100 \, kC = FC/\lambda \qquad \text{i.e.} \qquad F = - 100 \, k\lambda. \tag{7.8}$$

By substituting the respective values of k in equation (7.8), the blood flow per 100 g tissue may be found for white and grey matter.

Alternatively the ^{133}Xe may be administered by inhalation. The patient breathes on a closed circuit containing ^{133}Xe gas for five minutes and then breathes room air. The uptake of tracer by the brain and its subsequent clearance over 40 minutes is monitored by external scintillation detectors. The method has the advantage of being completely atraumatic. However, analysis of the data is more complex. Deconvolution techniques must be applied in order to obtain the shape of the curve that would have been obtained from a single carotid injection instead of the series of such injections effectively arising from each inhalation. Also there will be time for the ^{133}Xe to diffuse into extracerebral tissues and analysis of the clearance curve will be more complex as a result of these additional slowly clearing sources of activity.

7.4 Local Clearance Measurements

Local clearance measurements are based on following the disappearance of tracer locally deposited in tissue. In principle they are very similar to the measurement of cerebral blood flow by the ^{133}Xe injection technique described in § 7.3. The concept of clearance is the same as that used in the renal context, i.e. the imaginary volume of blood that would have contained the amount of tracer leaving 100 g of tissue if immediate equilibrium occurred. The mathematics of this process are those of the last part of § 7.2 and equation (7.8) applies, i.e.

$$\text{clearance} = -100\,k\lambda$$

where k is the fractional rate constant of an exponential clearance and λ is the tissue to blood partition coefficient. λ must be obtained for the tissue under study. The clearance of the activity is usually determined with a small radiation detector collimated to the size of the area of interest and connected to a ratemeter system. k is determined from the slope of the log(activity) versus time curve.

Two types of tracer have been used for the technique. The more common is the freely diffusible tracer, e.g. ^{133}Xe or ^{85}Kr. These have the advantages, due to their lypophilic nature, of virtually complete equilibration which is necessary for the theoretical reason outlined earlier, and their virtual elimination once they have left the measurement site which avoids the need for corrections for recirculation of the tracer. ^{133}Xe is more commonly used in conjunction with a solid scintillation detector because of the ease of external detection of its gamma radiation. ^{85}Kr, which is a beta emitter, has been employed in situations where the short range of the beta particles is helpful in limiting the depth of tissue contributing to the count rate. Surface measurements must then be made and these may be performed with an end window Geiger counter. The less common type of tracer has a restricted diffusibility and includes Na$^+$ and K$^+$ ions. These tracers ^{24}Na, ^{42}K and ^{43}K are hydrophilic ions which can only pass through the capillary membranes at water-filled pores. These are thought to occupy a very small part (approximately 0.1%) of the surface area. The rate of removal of these tracers is determined by capillary permeability in contrast to the diffusible tracers whose rate of removal is determined by blood flow. With these tracers only the rate constant k can be determined.

The technique is open to two difficulties. The first is that the introduction of the tracer must not distort the physical nature of the process under investigation. In practice this means that injected volumes must be small so that pressure effects are negligible. The second is that the tracer may enter other tissues and the clearance of the tracer may have to be corrected to take account of their handling of the tracer.

Local clearance techniques have been employed in a wide variety of tissues, e.g. muscle, skin, myocardium and the eye. For muscle blood flow 0.1 ml of ^{133}Xe in saline solution is injected into the muscle, and its clearance followed with a small solid scintillation detector with lead collimation 1–2 mm thick. This has been applied to exercise, when the lightweight detector is strapped to the leg over the gastrocnemius muscle (Lassen and Kampp 1965). In the normal subject the rate of clearance increased with exercise. In persons with claudication, the clearance is extremely slow during exercise and only begins to increase some two to three minutes after exercise has ceased. Skin blood has also been studied with ^{133}Xe in saline solution in a similar manner.

Coronary blood flow at different blood oxygen tensions has been studied using ^{133}Xe during cardiac catheterisation. ^{133}Xe in saline is administered as a bolus via a catheter inserted into one of the coronary arteries (Kenmure *et al* 1971). Venous samples are taken at 0.5 minute intervals for three minutes into sealed syringes which are later assayed for ^{133}Xe activity. If the net activities are plotted against time, the coronary blood flow can be determined from the fractional rate constant.

Choroidal blood in the eye has been determined experimentally in animals by administration of a small bolus of ^{85}Kr into the ciliac artery of the eye (Strang *et al* 1979). The clearance of the activity at the back of the eye has been monitored with either an end window Geiger counter or a planar solid state detector. The ^{85}Kr clearance curve is complex and has up to four components when analysed by curve stripping on a log(activity) versus time plot. The most rapid component corresponds to the choroidal blood flow and, with a normal flow of around 1100 ml min^{-1} per 100 g of tissue, is one of the highest blood flows in the body. The other slower components are related to retinal blood flow and diffusion from the humour and the sclera.

Blood flow has been considered more extensively by Rowan (1981).

8 Simulation of Physiological Systems

8.1 Introduction

The metabolism of a substance is the result of a number of inter-related dynamic processes. These include absorption, distribution, utilisation, degradation and excretion. Whilst the measurement of a single parameter as described in chapter 6 is diagnostic of a disorder, it may not be sufficient to indicate the cause of the disorder. This can only be done if a knowledge of the complete metabolic system is available. Attempting to simulate a system mathematically leads to the visualisation of the entire system and an understanding of the interaction of the component parts.

In order to model a physiological system, experimental data must be obtained following the stimulation of the system, usually in this instance by the addition of a suitable tracer. The tracer must clearly model the behaviour of the substance under investigation and not perturb this behaviour. The experimental data is then compared to the data predicted by the model, whose parameters are varied until the two sets of data agree as closely as possible. This process is usually done using the method of least squares (Whittaker and Robinson 1944, Berman *et al* 1962). In this method, the sum of the squared deviations between each data point and the corresponding theoretical value is minimised. The method of maximum likelihood (Sandor *et al* 1970) and Monte Carlo methods of predicting the theoretical values (Bartholomay 1968) have also been used. Digital computers play an important part in the iterative procedures necessary for the fitting of the data. Systematic differences between the predicted and experimental values may indicate that it is necessary to alter the composition of the model or the pattern of flows between its constituent parts. The experimental data can then be compared with the predictions of the new model to see if the goodness of fit has improved.

Certain general assumptions are almost always made about the system under investigation in addition to assuming that the addition of the tracer does not perturb the system. (The substance under investigation is often termed the 'tracee', i.e. the substance traced.) These are:

(a) the tracee is conserved throughout the process,
(b) the tracer is conserved throughout the process (after allowing for radioactive decay), and
(c) the system is in a steady state, i.e. the amount of tracee in each part of the system and the flows of tracee between each part remain constant.

In general two different types of model have been employed for simulating physiological processes. These are the deterministic type and the stochastic type.

A deterministic model is one in which analytic expressions are used to describe the exact behaviour of the tracer in each part of the system with time. Since the investigation of tracer kinetics usually involves the measurement of the rates of change of tracer content in different parts of the system, it is helpful to describe these changes by mathematical functions which are easily integrated to provide the time variations of the amount of material present. As shown in §6.3, the washout of tracer from a well mixed compartment is described by an exponential expression. This type of variation is conveniently integrated. Consequently compartmental analysis in terms of exponential functions has been very widely employed. This is especially true when measurements of the change in plasma concentration of tracer with time are involved. Two different approaches have been employed by workers using exponential analysis. Some analyse the variations in tracer concentration into the minimum number of exponentials using curve stripping (see figure 7.5) and arrive at the minimum number of compartments necessary. Others analyse their data using a comprehensive compartmental model which closely follows the physiology and anatomy of the system under investigation. The latter models tend to have more compartments than the former since the former approach lumps together compartments having similar temporal characteristics. Compartmental analysis has been examined extensively by Sheppard (1962) and Atkins (1969).

A few investigators (e.g. Wise 1978) have used other analytic functions, notably negative power functions of the form $y = a^{-bt}$, to describe their experimental data. However, these have proved difficult to interpret in physiological terms and have not been widely adopted.

The major drawback of compartmental analysis is the requirement for rapid mixing in each compartment. This is often only strictly true for the plasma compartment. Some investigators have preferred to adopt a more general approach with fewer implicit assumptions by assuming only that the behaviour of the system is determined in part by random processes described by probability functions. This is the stochastic approach. It is also appropriate to the system in which the analytic equations describing the system are non-linear or unknown. Few stochastic methods of analysis have been described in the literature. However, in one approach the system is seen as a collection of states with the radioactive particles moving from state to state with certain probabilities. The particle remains in each state for a random time, whose distribution is exponential. In a second, and more common theory, the stochastic method is used to examine the clearance of a tracer from a compartment, whose radioactive content is readily measured. This theory supposes that the particles leave the compartment, either never to return or to return after a stay of random length (the sojourn time) in another compartment.

In practice, the deterministic and stochastic approaches are not two distinct and separate methods of solving the problem of setting up a model which reasonably explains the observed behaviour of a system. Instead they tend to be used in a complementary fashion. When little is known about the system, the examination of a clearance curve using the stochastic model will reveal information about mean sojourn times outside the compartment of measurement in other compartments which can then be tentatively identified from other physiological data. This may enable the investigator to propose tentatively a more deterministic model against which the experimental data can be compared. As analysis proceeds and more experimental data are obtained this model may be refined to yield one which fits data in a wide variety of circumstances, both normal and pathological. Generally this should be the simplest model to fit the experimental results.

8.2   Compartmental Analysis

In deterministic modelling, a compartment may be an anatomical, physiological, chemical or physical subdivision of the system under investigation. As stated earlier, it is generally assumed that the tracer is rapidly mixed with the tracee on entry to the compartment so that the specific activity is uniform throughout the compartment at any moment. In this section the common compartmental models found in the literature will be outlined, together with their clinical application.

8.2.1   Single compartment models

The single compartment model shown in figure 8.1 has already been partly discussed in §6.3.1 in the situation where a bolus of radioactivity is added to visualise the flow through the compartment.

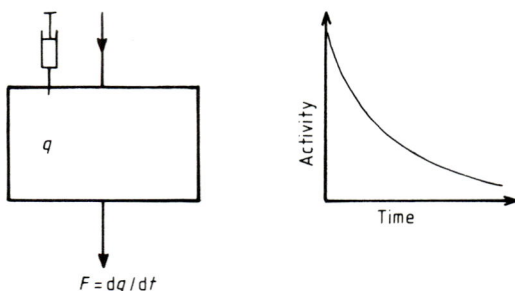

Figure 8.1 Single compartment model. The activity–time curve obtained when the tracer is delivered as a bolus is shown.

In this situation, $q(t)$ is the quantity of tracer in the compartment at time (t) and the outflow F is given by dq/dt. The fractional turnover k is the ratio of the outflow per unit time to the amount of tracer present, i.e.

$$k = \frac{-dq/dt}{q}. \tag{8.1}$$

Rearranging;

$$\frac{dq}{dt} = -kq. \tag{8.2}$$

This is the familiar expression for exponential loss. Integrating expression (8.2) yields the solution:

$$q = q_0 \, e^{-kt} \tag{8.3}$$

where q_0 is the amount of tracer present at time $t = 0$, when the tracer is added to the compartment.

This expression has already been met in a number of clinical situations, for example, the clearance of ^{59}Fe from the plasma to calculate the plasma iron transport rate (§6.3.2) and the local clearance of ^{133}Xe from a homogeneous tissue when a single blood flow is responsible for the removal of the gas (§7.4).

Alternatively the tracer can be introduced into the compartment at a constant rate, called the infusion rate f. If one starts with no radio-activity in the compartment, the radioactivity increases as the infusion proceeds. An equilibrium concentration is eventually reached at which the infusion rate equals the rate of loss of tracer in the outflow from the compartment. The rate of change of tracer content of the compartment will be given by

$$\frac{dq}{dt} = f - kq \tag{8.4}$$

using the same notation as before. When the system reaches a steady state $dq/dt = 0$, i.e.

$$f = kq_\infty \tag{8.5}$$

where q_∞ is the quantity of tracer in the compartment when the equilibrium concentration is reached. Eliminating f from expressions (8.4) and (8.5) yields the following expression

$$\frac{dq}{dt} = k(q_\infty - q). \tag{8.6}$$

Integrating this expression gives the time variation

$$q = q_\infty (1 - e^{-kt}). \tag{8.7}$$

This is the standard exponential build-up curve which is illustrated in figure 8.2.

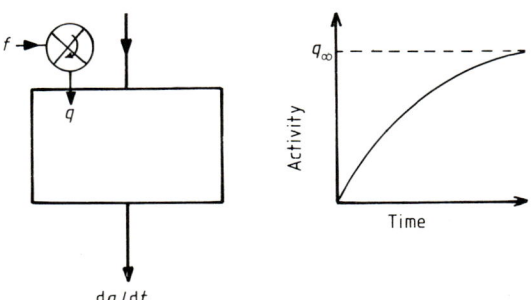

Figure 8.2 Single compartment model with constant infusion of tracer.

This situation is observed clinically with the build-up of the radio-activity in the blood during the measurement of renal clearance using the constant infusion method (§6.3.4). The plasma space is then the compartment. The outflow rate is set by the glomerular filtration rate or the effective renal plasma flow of the kidney and is proportional to the concentration of the tracer.

8.2.2   Two-compartment models

A small number of different configurations are possible with two compartments and the system as a whole may be closed or open to the outside. The three possible configurations will be considered in this section, a closed system with reversible transfer, and open systems with catenary and mamillary arrangements of the compartments.

The closed two-compartment model with reversible transfer is illustrated schematically in figure 8.3. The tracee moves in both directions between the compartments with no overall gain or loss of tracee. Suppose q_1 and q_2 are the tracer activities and k_{12} and k_{21} the fractional rate constants for compartments 1 and 2, respectively, then the rates of change of tracer content are given by the following expressions:

$$\frac{dq_1}{dt} = k_{21}q_2 - k_{12}q_1 \qquad \text{for compartment 1} \qquad (8.8)$$

$$\frac{dq_2}{dt} = k_{12}q_1 - k_{21}q_2 \qquad \text{for compartment 2.} \qquad (8.9)$$

Since there are no losses from the system, the tracer must be conserved, i.e.

$$q_1 + q_2 = \text{constant} = q_0$$

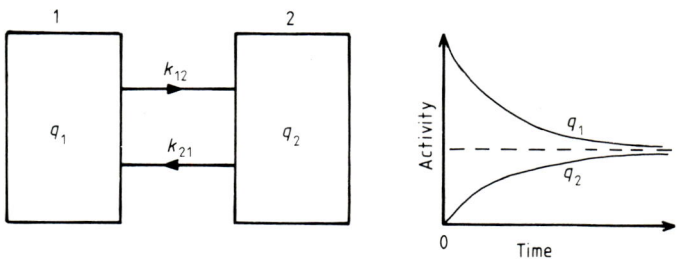

Figure 8.3   Closed two-compartment model with reversible transfer.

where q_0 is the added activity and

$$\frac{dq_1}{dt} = -\frac{dq_2}{dt}.$$

Suppose also that the bolus is administered to compartment 1 at time $t = 0$; then at $t = 0$,

$$q_1 = q_0 \qquad q_2 = 0$$

$$\frac{dq_1}{dt} = -k_{12}q_0 \qquad \frac{dq_2}{dt} = k_{12}q_0.$$

Integrating equations (8.8) and (8.9) and taking account of the initial conditions yields the following expressions for variations of tracer activity in the two compartments:

$$q_1 = q_0\left(1 - \frac{k_{12}}{k_{12} + k_{21}}\{1 - \exp[-(k_{12} + k_{21})t]\}\right) \qquad (8.10)$$

$$q_2 = q_0\frac{k_{12}}{k_{12} + k_{21}}\{1 - \exp[-(k_{12} + k_{21})t]\}. \qquad (8.11)$$

These variations are illustrated in figure 8.3. At equilibrium, the quantities of tracer in each compartment will be given by:

$$q_1 = q_0\frac{k_{21}}{k_{12} + k_{21}} \qquad (8.12)$$

and

$$q_2 = q_0\frac{k_{12}}{k_{12} + k_{21}} \qquad (8.13)$$

and the concentration of the tracer will be uniform throughout the system. If V_1 and V_2 are the volumes of compartments 1 and 2 respectively, then at equilibrium:

$$q_1/V_1 = q_2/V_2 \qquad \text{and} \qquad k_{12}q_1 = k_{21}q_2.$$

Eliminating q_1 and q_2 from these expressions gives:

$$k_{12}/k_{21} = V_2/V_1.$$

Equations (8.12) and (8.13) may also be expressed as:

$$q_1 = \frac{V_1}{V_1 + V_2}q_0 \qquad \text{and} \qquad q_2 = \frac{V_2}{V_1 + V_2}q_0,$$

The open two-compartment catenary model is illustrated in figure 8.4. Suppose the fractional exit rate constants are k_{12} and k_{20} for compartments 1 and 2 respectively, then for compartment 1, the rate of change of activity is given by:

$$\frac{dq_1}{dt} = -k_{12}q_1.$$

This is the equation for exponential loss described in equation (8.2), which gives on integration

$$q_1 = q_0\, e^{-k_{12}t} \tag{8.14}$$

where q_0 is the initial bolus activity. The rate of change of tracer activity in compartment 2 is given by:

$$\frac{dq_2}{dt} = k_{12}q_1 - k_{20}q_2. \tag{8.15}$$

Using equation (8.14) to eliminate q_1 from equation (8.15) and integrating, one obtains the equation:

$$q_2 = q_0\frac{k_{12}}{k_{12} - k_{20}}\,(e^{-k_{20}t} - e^{-k_{12}t}). \tag{8.16}$$

In the steady state, the flow of tracee through the system will be constant, i.e.

$$k_{12}V_1 = k_{20}V_2$$

where V_1 and V_2 are the volumes of compartments 1 and 2 respectively. Equation (8.16) can also be expressed as:

$$q_2 = q_0\frac{V_2}{V_2 - V_1}\,(e^{-k_{20}t} - e^{-k_{12}t}).$$

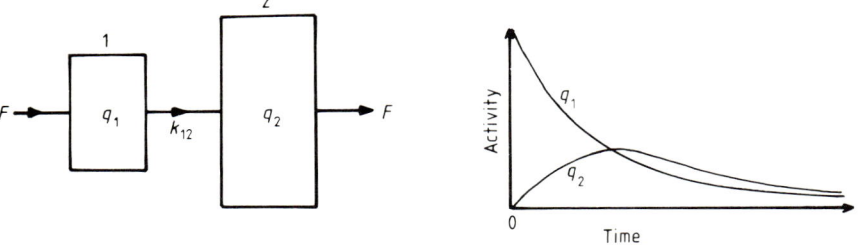

Figure 8.4   Open two-compartment catenary model.

Equations (8.14) and (8.16) are illustrated graphically in figure 8.4. It can be seen that the activity in compartment 2 rises until it reaches a maximum at which moment the tracer concentrations in the two compartments are equal. Thereafter the activity falls, but lags behind the activity level in compartment 1.

These equations are of course the same as those describing the activity variations with time of the parent and daughter radionuclides in the radionuclide generator (§1.5). In the clinical field, this model has been used most often to describe processes in which tracer is added to visualise the conversion of one material to another.

The catenary model has been discussed more extensively by Berman and Schoenfeld (1956). Additional compartments can, of course, be added to extend the range of the model.

The open two-compartment mamillary model is illustrated in figure 8.5. In mamillary models compartments are linked to a central compartment which is open; this shows the simplest arrangement with one peripheral compartment. Suppose the tracer bolus is added to compartment 1. Using the same nomenclature as before, the rates of change of tracer activity in the two compartments are given by the following expressions.

$$\frac{dq_1}{dt} = -k_{10}q_1 - k_{12}q_1 + k_{21}q_2 \tag{8.17}$$

and

$$\frac{dq_2}{dt} = k_{12}q_1 - k_{21}q_2. \tag{8.18}$$

k_{10} is the fractional excretion rate from compartment 1. Taking account of the starting conditions at $t = 0$ when

$$q_1 = q_0 \qquad q_2 = 0$$

and

$$dq_1/dt = -(k_{10} + k_{12})q_1 \qquad dq_2/dt = k_{12}q_0,$$

equations (8.17) and (8.18) may be integrated. Then:

$$q_1 = q_0 \left(\frac{k_{21} - a_1}{a_2 - a_1} e^{-a_1 t} + \frac{k_{21} - a_2}{a_1 - a_2} e^{-a_2 t} \right) \tag{8.19}$$

$$q_2 = \frac{q_0 k_{12}}{a_2 - a_1} (e^{-a_1 t} - e^{-a_2 t}) \tag{8.20}$$

where $a_1a_2 = k_{10}k_{21}$ and $a_1 + a_2 = k_{10} + k_{21} + k_{12}$. The variations of q_1 and q_2 with time are shown in figure 8.5, together with the total activity $q_1 + q_2$. The rate constants can be obtained by plotting the variation of q_1 on a log–linear plot and analysing the curve into two linear components using curve stripping.

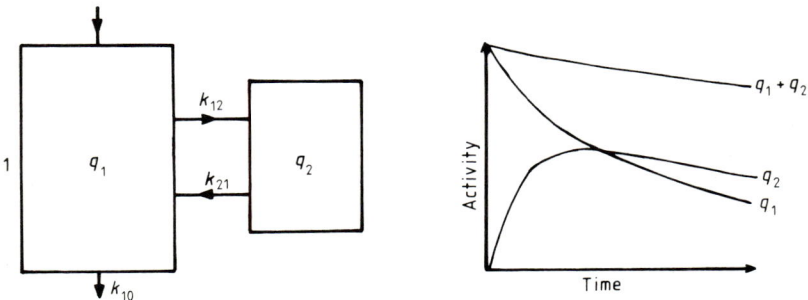

Figure 8.5   Open two-compartment mamillary model.

As in the catenary model above, there is a time when the concentrations in compartments 1 and 2 are equal and this is also the time when the concentration in compartment 2 is maximal, i.e. when

$$\frac{dq_2}{dt} = 0.$$

Then, from equation (8.18),

$$k_{12}q_1 = k_{21}q_2. \tag{8.21}$$

Also, in steady state conditions, the flows of tracee into and out of compartment 2 must balance, i.e.

$$k_{12}V_1 = k_{21}V_2 \tag{8.22}$$

where V_1 and V_2 are the volumes of compartments 1 and 2 respectively. Combining equations (8.21) and (8.22) yields the expression:

$$q_2/q_1 = V_2/V_1,$$

i.e. the ratio of tracer activities equals the ratio of compartment volumes at the time of peak activity in compartment 2.

The excretion rate of the tracer dq_0/dt is given by the equation:

$$\frac{dq_0}{dt} = -k_{10}q_1,$$

i.e.

$$k_{10} = \frac{-dq_0/dt}{q_1}.$$

The excretion rate constant is therefore given by the rate of excretion of activity divided by the activity in compartment 1.

The mamillary model has been considered more extensively by Sheppard and Householder (1951) and Matthews (1957).

This model has been used extensively to study the metabolism of plasma proteins, compartment 1 being the plasma and compartment 2 the extravascular space. It has also been used to study the trapping of the pertechnetate ion ($^{99m}TcO_4^-$) by the thyroid gland. Compartment 1 is the plasma and compartment 2 is the gland, k_{12} is the clearance rate and k_{21} is the leakage rate of pertechnetate back into the plasma. An extension of the model has been used to study the uptake of iodide (I^-) by the thyroid gland (Hilditch *et al* 1981). A third compartment is added as shown in figure 8.6 which is fed by an irreversible flow k_2 from compartment 2. In this model, compartment 1 represents the plasma, compartment 2 represents the trapping of inorganic iodide in the gland and compartment 3 the iodide within the gland which has become organically bound as part of hormone synthesis.

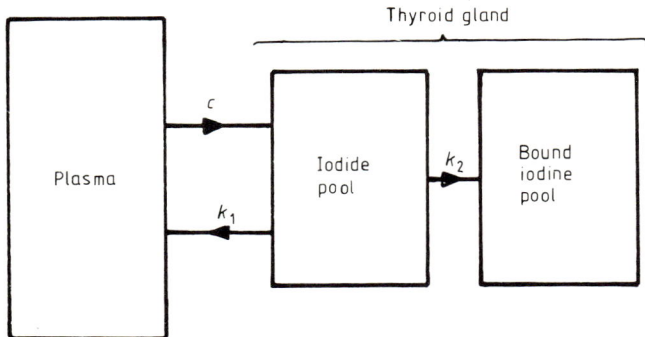

Figure 8.6 Model describing handling of iodide by the thyroid gland. c is the clearance rate of iodide from the blood, k_1 is the leak rate back into the blood and k_2 is the binding rate of iodide.

8.2.3 Three-compartment models

In general, curve stripping on a log–linear plot into exponential components can only be done for a maximum of three components. Experimental errors in the data usually render further stripping inaccurate.

For a closed system such as that shown in figure 8.7, the activity–time curves for each compartment will have the form

$$q = a_1 e^{-b_1 t} + a_2 e^{-b_2 t} + c$$

where a_1, a_2, b_1, b_2 and c are constants.

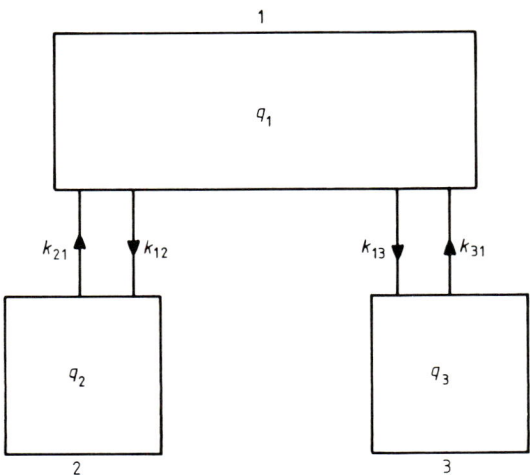

Figure 8.7   Closed three-compartment model.

Open systems like that shown in figure 8.8 have activity–time curves of the form:

$$q = a_1 e^{-b_1 t} + a_2 e^{-b_2 t} + a_3 e^{-b_3 t}.$$

The model shown in figure 8.8 is often used in a simplified form in which it is assumed that the outflows from compartments 2 and 3 are negligible by comparison with the inflows, i.e.

$$k_{20} = k_{30} = 0.$$

Under these circumstances, the rates of change of activity in the three compartments are given by the equations:

$$\frac{dq_1}{dt} = k_{12}q_1 - k_{13}q_1, \tag{8.23}$$

$$\frac{dq_2}{dt} = k_{12}q_1, \tag{8.24}$$

$$\frac{dq_3}{dt} = k_{13}q_1. \tag{8.25}$$

The activity–time variation for compartments is readily obtained by integration, i.e.

$$q_1 = q_0 \exp[-(k_{12} + k_{13})t]. \tag{8.26}$$

Using expression (8.26) to eliminate q_1, expressions (8.24) and (8.25) become:

$$\frac{dq_2}{dt} = k_{12}q_0 \exp[-(k_{12} + k_{13})t]$$

$$\frac{dq_3}{dt} = k_{13}q_0 \exp[-(k_{12} + k_{13})t].$$

Integration of these expressions gives the following equations for q_2 and q_3:

$$q_2 = \frac{k_{12}q_0}{k_{12} + k_{13}} \{1 - \exp[-(k_{12} + k_{13})t]\}$$

$$q_3 = \frac{k_{13}q_0}{k_{12} + k_{13}} \{1 - \exp[-(k_{12} + k_{13})t]\}.$$

This model has been used to describe the clearance of iodide from the plasma into the thyroid gland and the kidneys.

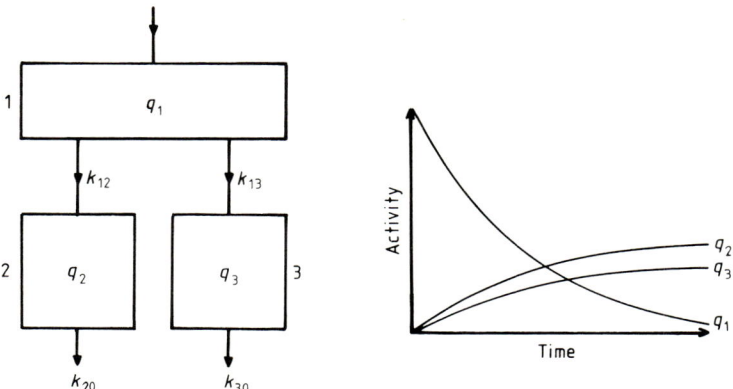

Figure 8.8   Open three-compartment model.

8.3   Stochastic Methods

The mathematics required for the stochastic analysis of multicompartment systems is complex and not appropriate to the introductory nature of this book.

Marsaglia (1963) has used matrix methods to solve the equations for the model in which the compartments are visualised as a collection of states with radioactive particles moving between them with finite probabilities related to the flow rates between compartments. The particles are assumed to remain in each state for a random time, whose distribution function is exponential.

Marsaglia has also examined the stochastic approach using the concept of 'sojourn' of tracer outside the compartment being measured. Suppose a radioactive tracer is added to the compartment at time $t = 0$ and the radioactivity is measured at intervals. Suppose the variation in tracer activity with time can be described by the function $q(t)$. Over a period of time, radioactivity will leave the compartment. Some will not return to the compartment but some will return after a random sojourn time in another compartment. Suppose that $s(x)$ described the distribution of sojourn times over all other compartments, then the rate of change of activity can be described by the expression

$$\frac{dq}{dt} = -hq(t) + h \int_0^t q(x)s(t-x) \, dx$$

where h is the initial rate of change of tracer activity in the measurement compartment immediately following tracer administration, i.e.

$$h = \frac{dq}{dt}\bigg|_{t=0}.$$

The first term $-hq(t)$ therefore represents the flow from the compartment. The second term represents the reflux from all other compartments to which the tracer flows and can return. It represents the convolution of the activity entering these compartments at previous times with the proportion returned at each time interval.

Marsaglia has shown that if $q(t)$ is a linear combination of exponentials, then $s(x)$ is always a linear combination with one fewer exponentials. $s(x)$ has been calculated in the following cases.

(a) Suppose

$$q(t) = a_1 e^{-k_1 t} + a_2 e^{-k_2 t}$$

with

$$q(0) = 1 \quad \text{and} \quad dq/dt \, (0) = -h,$$

then

$$s(x) = c\,e^{-bx}$$

where

$$b = k_1 + k_2 - h \qquad \text{and} \qquad c = -[(h - k_1)(h - k_2)]/h.$$

(b) Suppose

$$q(t) = a_1\,e^{-k_1 t} + a_2\,e^{-k_2 t} + a_3\,e^{-k_3 t}$$

then

$$s(x) = c_1\,e^{-b_1 x} + c_2\,e^{-b_2 x}.$$

b_1 and b_2 are given by the roots of the quadratic equation:

$$x^2 - (k_1 + k_2 + k_3 - h)x + a_1 k_2 k_3 + a_2 k_1 k_3 + a_3 k_1 k_2 = 0.$$

c_1 and c_2 satisfy the two linear equations:

$$c_1 + c_2 = b_1 + b_2 - \frac{k_1 + k_2 + k_1 k_3 + k_2 k_3 - b_1 b_2}{h}$$

$$b_2 c_1 + b_1 c_2 = b_1 b_2 - \frac{k_1 k_2 k_3}{h}.$$

The integral of $s(x)$ will give the area beneath the reflux function and the probability of the tracer returning to the compartment under observation. If $s(x)$ is divided by this probability, the density function of the sojourn time is obtained

This type of analysis is helpful in determining the number of compartments to which a tracer may be going from the plasma pool following dose administration. The coefficients b, b_1, b_2, etc, provide an indication of the turnover rate in each compartment and the coefficients c, c_1, c_2, etc, indicate the proportion of the flow going to each of these compartments. This work by Marsaglia was first applied in ferro-kinetics to examine the clearance of ^{59}Fe from plasma (Cook *et al* 1970). The reflux of ^{59}Fe in the plasma clearance curve was found to contain two components; a smaller short term one which has been associated with the exchange of iron in lymphatic tissue and a larger long term one which has been associated with the return of iron from ineffective erythropoiesis in bone marrow. This second component is increased in magnitude in some anaemias.

The concept of sojourn time for radioactivity in physiological compartments has also been used by Bergner (1961, 1962, 1964) in his extensive analysis of tracer dynamics. However, Bergner used a different definition from Marsaglia and applied the term to the compartment under observation. Bergner's analysis requires a good mathematical knowledge and reference should be made to his original papers for the details. However, one of his theorems has been presented in a more convenient form for practical use by Orr and Gillespie (1968). It is a useful theorem since it enables measurements to be made of physiological quantities which are not directly accessible to measurement.

The theorem has been termed the 'occupancy principle' by Orr and Gillespie and states that the ratio of occupancy θ, to the capacity C, is the same for all compartments of the system and is equal to the reciprocal of the entry flow, F, i.e.

$$\frac{\theta_1}{C_1} = \frac{\theta_2}{C_2} = \frac{\theta_i}{C_i} = \frac{\theta_n}{C_n} = \frac{1}{F} \tag{8.27}$$

where the subscript denotes the compartment number. The occupancy for any compartment is defined as the total integral with respect to time of the fraction of the tracer $f(t)$ in that part of the system, i.e.

$$\theta = \int_0^\infty f(t) \, dt.$$

This is the area beneath the activity–time variation for the compartment concerned (figure 8.9). The capacity is the quantity of non-radioactive material under study in the compartment which has the same form as the tracer. Capacity has the same meaning as the terms 'amount of tracer', 'amount of mother substance' and 'pool size'.

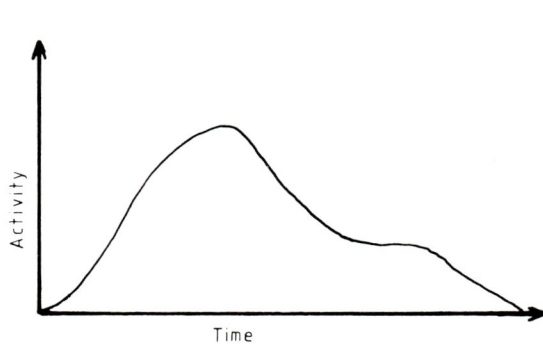

Figure 8.9   Occupancy. The occupancy is given by the area beneath the activity–time curve.

The principle can be proved as follows. Suppose at time $t = 0$ in the interval dt a quantity of the tracee given by $F dt$ entered the system along with the tracer. A fraction $f(t)$ of the tracer is then contained in the defined part of the system at time t. Since the behaviour of the tracer and the tracee are identical, the same fraction $f(t)$ of the stable material must also be present in the defined part at time t. The total amount of material C in the defined part of the system is therefore the sum of all the material that entered the system from $t = 0$ to ∞, i.e.

$$C = \int_0^\infty F \, dt \, f(t)$$

$$= F \int_0^\infty f(t) \, dt = F\theta.$$

Application of the principle is straightforward for compartments in the catenary arrangement shown in figure 8.10 when one of the compartments is accessible. Suppose compartment 1 is the plasma. The plasma occupancy can be measured by calculating the area beneath the clearance curve. This can either be done mathematically using analytic expressions which describe $f(t)$, or by numerical integration, sometimes with the addition of the area beneath a final analytic function to give the area to infinite time. The capacity of the tracer substance in the plasma can be measured by biochemical techniques. If compartment 2 is less accessible, one has difficulty in measuring the amount of stable material present. However, the occupancy might be measurable through external counting. Substitution of these three measured quantities in expression (8.27) enables the capacity or amount of tracee in compartment 2 to be calculated.

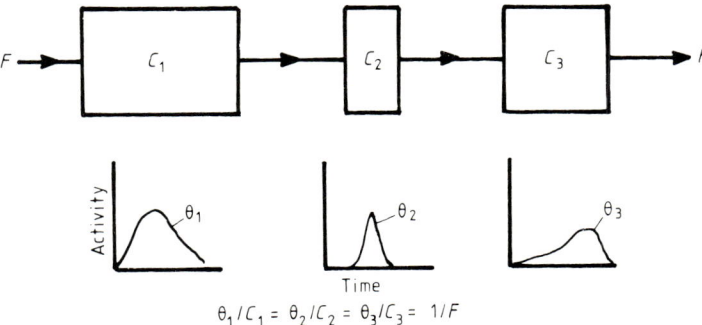

Figure 8.10 Application of the occupancy principle to a catenary model.

   Whilst the principle appears to be extremely simple (equation (8.27)), care is necessary for its proper application. Zierler (1969) mistakenly identified the tracer sojourn time (Bergner 1964), or occupancy, with mean transit time. This is only true for an entire system or for a compartment in the system for which there is no recirculation of its outflow. It cannot generally be assumed to be correct for a compartment within the system (Perl and Chinard 1969). The important consideration in applying the occupancy principle is that the tracer must enter along with the entry flow and whilst recirculation is allowed within the system, no tracer should leave the system and re-enter with the entry flow. This is illustrated in figure 8.11. In measuring $f(t)$ by sampling the activity in compartment 1, one is automatically taking account of the recirculation of tracer from compartments 2 and 3. The part of the system to which the occupancy principle is applied is therefore the system inside the rectangular box and the flow F into this rectangle. In this situation it can clearly be seen that the occupancy given by the area beneath the activity–time curve for compartment 1 can be considerably greater than the mean transit time (given by the area beneath the activity–time curve for activity in the first passage only, if this could be measured separately) due to the effects of recirculated tracer from compartments 1 and 2. Conversely the flow F into the defined part of the system can be

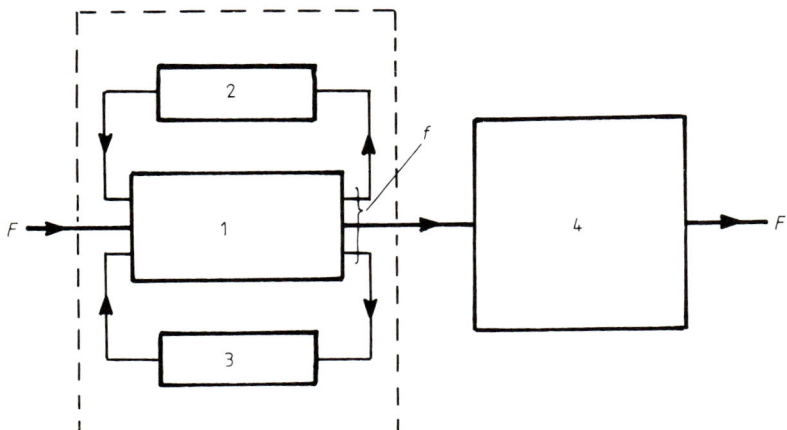

Figure 8.11 Application of the occupancy principle to a system with recirculation. Since the occupancy of compartment 1 includes the fluxes from compartments 2 and 3, the system to which the principle applies is contained within the rectangle. The flow F in the principle is the flow into the rectangle.

considerably less than the total flow f into compartment 1. This point is well illustrated by reference to ferro-kinetics where compartments 1, 2, 3 and 4 are the plasma, lymphatic system, ineffective erythropoiesis and red cells. The flow F then represents the true flow of iron into red cells for the synthesis of haemoglobin, i.e. effective erythropoiesis. The total flow f from compartment 1 corresponds to the plasma iron transport rate (§6.3). This includes the recirculation flows to compartments 2 and 3 and is not the true flow of iron into effective erythropoiesis. As stated earlier, f can be considerably in excess of F if appreciably ineffective erythropoiesis is present, as happens in certain anaemias. This application to ferro-kinetics has been used to determine red cell lifespan using ^{59}Fe (Dagg *et al* 1972, Hutcheon *et al* 1977). This has considerable scientific advantages over the ^{51}Cr method since the radio-iron is an intrinsic label and does not have the inaccuracies associated with elution of the ^{51}Cr label described in §6.33.

The occupancy principle has been employed in a number of other clinical contexts, notably thyroid hormone metabolism (Harland and Orr 1969), thyroid iodine content (Riviere *et al* 1969) and determination of exchangeable mass (Bergner 1965). It has been discussed more extensively by Orr (1975).

9 Radioimmunoassay and Related Techniques

9.1 Introduction

Radioimmunoassay (RIA) is a laboratory technique for measuring the levels of substances which have low concentrations in the blood, e.g. hormones. It is extremely sensitive and can sometimes detect concentrations of 10^{-15} moll^{-1}. This is between 10^3 and 10^6 times more sensitive than colorimetric measurements. The technique was introduced in 1959 by Berson and Yalow, who developed a RIA for insulin, and independently by Ekins, who developed an assay for the thyroid hormone, thyroxine. Since then the range of substances which can be assayed has expanded tremendously, as indicated by the length of the list in table 9.1. The list of hormones alone shows that modern endocrinological diagnosis could hardly function without the technique. Similarly the volume of testing has also increased greatly; hospitals in the United Kingdom report a ten-fold increase over the past ten years.

In radioimmunoassay, the radioactivity is processed *in vitro* together with biological substances. This is a somewhat different type of clinical application to those described so far in this book because the patient is remote from the radioactivity. This is a desirable feature of radioimmunoassay since it avoids giving the patient a radiation dose, however small.

9.2 Principle of Operation

In outline, radioimmunoassay entails the interaction of three separate entities. These are:

(a) a plasma sample from the patient containing the substance whose concentration is to be measured,

(b) the same substance with a radioactive label, and

(c) receptors which bind only the substance of interest.

154

Table 9.1 Compounds assayed by radioimmunoassay.

Hormones	Thyroxine (T4)	Tri-iodothyronine (T3)
	Thyroid stimulating hormone (TSH)	Grown-hormone (HGH)
	Luteinising hormone (LH)	Follicle-stimulating hormone (FSH)
	Prolactin	Adrenocorticotropic hormone (ACTH)
	Cortisol	Oestradiol
	Progesterone	Testosterone
	Oestriol	Chorionic gonadotropin (HCG)
	Aldosterone	Parathormone (PTH)
	Calcitonin	Insulin
	Gastrin	
Proteins	Placental lactogen (HPL)	Carcinoembryonic antigen
	Alpha fetroprotein	Thyroid binding globulin (TBG)
	Ferritin	Immunoglobulin (IGE)
Vitamins	Vitamin B$_{12}$	Folic acid
Drugs	Digoxin	Digitoxin
	Phenobarbitone	Phenytoin
	Paraquat	Methotrexate
	LSD	Cannabis
Others	Renin	Hepatitis B Virus
	Cyclic-AMP	

The number of receptor molecules is made equal to the number of molecules of the radioactively labelled substance. During the incubation period after the three entities have been added together, the molecules of the substance from the patient compete with the labelled molecules to form complexes with the receptors. At the end of the incubation period the ratio of labelled complexes to unlabelled complexes reflects the concentration of labelled substances to patient substance; the greater the concentration of the substance in the patient's plasma, the less the number of labelled complexes. The proportion of labelled complexes can be measured by separating the complexes from the unbound substance, both labelled and unlabelled, and counting the radioactivity present. The more of the substance in the plasma sample, the lower the radioactivity. This procedure is illustrated schematically in figure 9.1 for two samples of plasma in which the concentrations of the substance to be assayed differ by a factor of three. This shows how the measured radioactivity is inversely related to the concentration of the substance to be measured.

In true radioimmunoassay, the binding receptor molecule is an antibody. This results in a very specific assay due to the avidity between the substance under measurement (the antigen) and the antibody. Antibodies have such high specificity that they can distinguish between two peptides which differ by as little as one atom, e.g. thyroxine and tri-iodothyronine. A requirement of radioimmunoassay is therefore the availability of an antiserum to the compound to be measured.

However the technique is not limited to the use of antiserum; any agent that has the property of binding to a specific molecule may be used. Examples include thyroid binding globulin and transcortin. Assays using proteins such as these are termed competitive protein binding assays. Such assays usually lack the specificity of radioimmunoassay.

The term 'radioimmunoassay' is now often applied to the whole field, although not all assays rely on antisera. Other terms more common in the past are saturation analysis, competitive radioassay, competitive binding assay and displacement analysis. None of these terms is universally appropriate. Thorell and Larson (1978) have adopted the general name 'radioligand assays' because all the techniques involve the binding of a radioactive substance. The presence of the radioactive label is the second reason for the very successful introduction of the technique. This enables the extremely sensitive methods of detecting radioactivity to be employed and endows the technique with sensitivity as well as specificity. This was facilitated by the development in the early 1960s of the

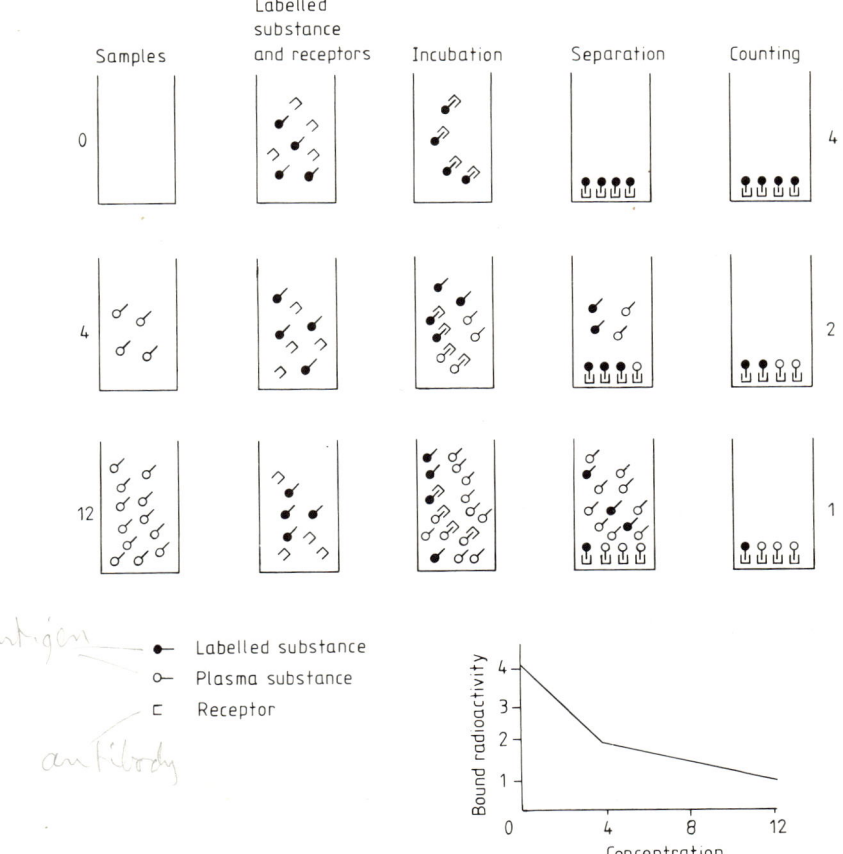

Figure 9.1 Principle of radioimmunoassay.

chloramine-T method for labelling proteins with high activities of radioiodine.

9.3 Practical Considerations

9.3.1 Production of antisera

To produce the antiserum, the antigen is injected subcutaneously or intraperitoneally into an experimental animal, usually the rabbit, together with an adjuvant to stimulate the immune system. Large molecules with molecular weights over 10 000 are naturally antigenic and cause the production of antibodies. Small molecules such as thyroid and steroid hormones and drugs are not antigenic. These are made antigenic by a process called 'haptenisation' in which the molecule is linked to a large molecule like albumin. If after further injections of the antigen over a period of weeks, the correct antibody level is reached, the animal is bled to produce an antiserum. The antiserum is then tested against molecules other than the substance of interest (antigen) to determine its cross reactivity. This must be low for other substances that might be expected in the plasma samples from patients.

9.3.2   Radioactive label

The choice of a radioactive label for the antigen will depend on its specific activity, its physical half-life and biological and chemical stability. Various radionuclides have been employed; ^{3}H, ^{51}Cr, ^{57}Co, ^{59}Fe, ^{75}Se, ^{125}I and ^{131}I. ^{125}I is generally the label of choice. Its half-life of 60 days enables the labelled antigen to be kept for two to three months before use. Its soft 30 keV x-radiation minimises radiolytic decomposition. The availability of high specific activity iodide preparations enables the production of antigen with high specific activity and sensitive assays. The chloramine-T method is the most common one for labelling with ^{125}I.

9.3.3   Separation techniques

A wide range of techniques has been employed to separate the bound and free material after incubation. These include electrophoresis, chromatography, gel filtration, ion exchange, charcoal–dextran absorption, double antibody precipitation and solid phase methods. The last three methods are the commonest.

The charcoal–dextran method is used for small molecules. A suspension of fine charcoal powder and dextran is added to the mixture. The free antigen, labelled and unlabelled, binds to the charcoal and is removed by centrifugation leaving the bound material in solution.

The double antibody technique is used for larger molecules. A second antiserum raised in a different species of animal from the first antiserum is added to the mixture and a very large antibody complex results. This can be precipitated from the solution for counting.

In solid phase methods, the first or second antibody is bound to sepharose or cellulose beads. This yields an excellent separation of the free and bound materials.

9.3.4 Counting techniques

In general use, plasma samples from patients will be batched to be processed together along with standards which relate the net count rates in each sample to the plasma concentrations of the substance being measured. In a busy centre, up to 300 samples may be grouped together for efficient processing.

Counting of the activity of the samples may then take place on an automatic gamma counter (§ 3.4.2) for gamma emitting labels or on a liquid scintillation counter (§ 4.1.6) for a ^{3}H labelled assay. This can take a considerable amount of time for a large number of samples because of the interval whilst each new sample enters the scintillation detector.

An alternative method of counting samples with ^{125}I or ^{57}Co labels has been introduced. A new automatic gamma counter has been developed which contains twelve or sixteen separate well type scintillation detectors. The reaction tubes containing the incubation mixture are loaded into trays which fit directly into the array of detectors. In this way, up to sixteen samples may be loaded simultaneously and the samples counted in parallel for a preset period. The results from all the detectors are recorded by a printer. Since counting times are short, the throughput of one of these counters can be considerable. An advantage of this type of counter is that there are no moving parts to reduce reliability. One example of this type of counter is shown in figure 9.2.

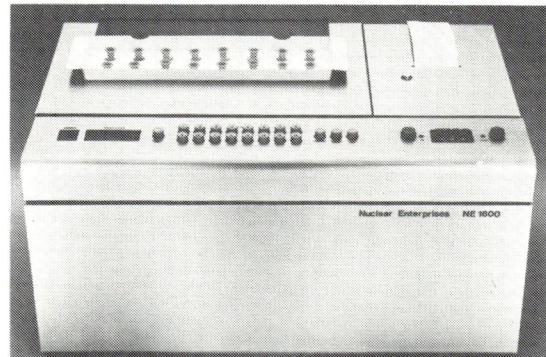

Figure 9.2 Multi-detector gamma counter. The tray holding sixteen samples for simultaneous counting can be seen at the rear left. The display showing the counts in each detector channel and the elapsed time together with their selection pushbuttons can be seen front left. A list printer for results can be seen right. (Reproduced by kind permission of EMI Nuclear Enterprises Limited.)

Both types of gamma counter are available with microprocessors for the data processing of results. The counts from a set of standards with a range of concentrations of the substance of interest may be entered and a calibration curve derived. Counts from individual samples may then be compared with the curve to derive the plasma concentration of the substance.

The cost of the radionuclide counter is one of the drawbacks of establishing a radioimmunoassay service. This has been cited as one of the reasons why radioimmunoassay will be superseded by other techniques such as immunofluorescence and enzyme linked radioimmunoassays (such as EMIT and ELIZA), which have lower capital costs. They also avoid the use of radioactivity, though they do bring other hazards. At the present time, these new techniques are relatively untried and their impact on radioimmunoassay has still to be measured.

9.4   Applications

The list in table 9.1 shows that radioimmunoassay has a very valuable role in the estimation of hormone levels in the blood and therefore as a diagnostic tool in endocrinology. Radioimmunoassay is also widely employed to measure proteins, vitamins and drugs. Many of these assays are available in the form of kits from commercial manufacturers. A kit comprises the antiserum to the substance to be measured, the radioactively labelled substance, a mechanism for separating the receptor bound activity from the unbound activity and standards of unlabelled material.

A survey by Smith (1980) has shown that there are 307 different kits offered by 22 companies in the United Kingdom. A critical awareness of the principles of radioimmunoassay and its drawbacks must be exercised in using a commercial kit. They differ in their sensitivity, specificity, incubation technique and mode of separation.

Appendix: Radionuclides in Clinical Use

Radio-nuclide	Half-life	Decay† process	Principal radiation (MeV)		Production method	Usage	Section reference
			e	x or γ			
^{3}H	12.26 yr	e⁻	0.018	—	^{6}Li(n, α)^{3}H	Whole body water / Biochemical research	5.4.1
^{11}C	20.3 min	e⁺	0.97	0.511	^{10}B(d, n)^{11}C	Physiological research	—
^{14}C	5570 yr	e⁻	0.155	—	^{14}N(n, p)^{14}C	Breath tests	6.2.6
^{13}N	10.0 min	e⁺	1.20	0.511	^{12}C(d, n)^{13}N	Physiological research	—
^{15}O	2.05 min	e⁺	1.74	0.511	^{14}N(d, n)^{15}O	Physiological research	—
^{18}F	110 min	e⁺, EC	0.63	0.511	^{16}O(α, pn)^{18}F	Cancer research	—
^{24}Na	2.58 yr	e⁻	1.39	1.37 / 2.75	^{23}Na(n, γ)^{24}Na	Exchangeable sodium	5.4.3
^{32}P	14.45 d	e⁻	1.71	—	^{31}P(n, γ)^{32}P	Therapy of polycythaemia	—
^{35}S	87 d	e⁻	0.167	—	^{35}Cl(n, p)^{35}S	Drug research	—
^{36}Cl	310 000 yr	e⁻, EC	0.71	—	^{35}Cl(n, γ)^{36}Cl	Physiological research	5.4.2
^{42}K	12.5 h	e⁻	2.0 / 3.6	1.53	^{41}K(n, γ)^{42}K	Exchangeable potassium	5.4.3
^{43}K	22 h	e⁻	0.83	0.37 / 0.61	^{40}A(α, p)^{43}K	Exchangeable potassium	5.4.3
^{45}Ca	165 d	e⁻	0.25	—	^{44}Ca(n, γ)^{45}Ca	Calcium kinetics	6.2.1
^{47}Ca →	4.53 d	e⁻	0.69 / 2.00	1.31	^{46}Ca(n, γ)^{47}Ca	Calcium kinetics	6.2.1
^{47}Sc	3.43 d	e⁻	0.44 / 0.60	0.16	—	—	—

Radio-nuclide	Half-life	Decay† process	Principal radiation (MeV)		Production method	Usage	Section reference
			e	x or γ			
^{51}Cr	27.8 d	EC	—	0.322	^{50}Cr(n, γ)^{51}Cr	Red cell labelling Glomerular filtration rate	5.3, 6.3.3 6.3.4
^{52}Fe	8.3 h	e$^+$, EC	0.81	0.511	^{52}Cr(α, 4n)^{52}Fe	Bone marrow imaging	—
^{55}Fe	2.7 yr	EC	—	0.006	^{54}Fe(n, γ)^{55}Fe	Ferrokinetics	6.2.1
^{59}Fe	45 d	e$^-$	0.27 0.46	1.10 1.29	^{58}Fe(n, γ)^{59}Fe	Ferrokinetics	6.2.1, 6.3.2
^{57}Co	267 d	EC	—	0.122	^{60}Ni(p, α)^{57}Co	Vitamin B$_{12}$ absorption	6.2.3, 6.2.5
^{58}Co	71 d	e$^+$, EC	0.49	0.511 0.81	^{58}Ni(n, p)^{58}Co	Vitamin B$_{12}$ absorption	6.2.3, 6.2.5
^{65}Zn	245 d	e$^+$, EC	0.33	0.511 1.11	^{64}Zn(n, γ)^{65}Zn	Physiological research	—
^{67}Ga	78 h	EC	—	0.18 0.30	^{65}Cu(α, 2n)^{67}Ga	Location of neoplasms and abscesses	6.4.2
^{75}Se	120 d	EC	—	0.14 0.27	^{74}Se(n, γ)^{75}Se	Imaging of the pancreas and adrenal glands	6.4.2
^{81}Rb	4.5 h	e$^+$, EC	—	0.511	^{79}Br(α, 2n)^{81}Rb	Radionuclide generator	1.5
→ ^{81m}Kr	13.5 s	IT	—	0.190	—	Lung function studies	6.4.2
^{77}Br	58 h	e$^+$, EC	—	0.520	^{75}As(α, 2n)^{77}Br	Extracellular water	5.4.2
^{82}Br	35 h	e$^-$	0.44	0.55 0.62 0.78	^{81}Br(n, γ)^{82}Br	Extracellular water	5.4.2
^{90}Y	64.4 h	e$^-$	2.27	—	^{89}Y(n, γ)^{90}Y	Treatment of arthritic joints	—
^{99}Mo	67 h	e$^-$	0.45 1.23	0.74	^{98}Mo(n, γ)^{99}Mo U(n, f) → ^{99}Mo	Radionuclide generator	1.5
→ ^{99m}Tc	6 h	IT	—	0.141		Organ imaging (table 6.1)	6.4.2

Nuclide	Half-life	†			Production	Application	Reference
111In	2.8 d	EC	—	0.17	109Ag(α, 2n)111In	White cell labelling	—
				0.25		Imaging of cerebrospinal fluid	6.4.2
113Sn → 113mIn	118 d	EC	—	0.26	112Sn(n, γ)113Sn	Radionuclide generator	1.5
	104 m	IT	—	0.39	—		
123I	13.3 h	EC	—	0.16	121Sb(α, 2n)123I 127I(p, 5n)123Xe→ 123I	Cardiac output	7.2
						Thyroid studies	6.4.2, 6.4.3
						Renal studies	6.4.2, 6.4.4
125I	60 d	EC	—	0.035	124Xe(n, γ)125Xe→ 125I	Radioimmunoassay	9
						Plasma volume	5.2
						Effective renal plasma flow	6.3.4
						Deep vein thrombosis	6.4.5
131I	8.1 d	e⁻	0.61	0.36	130Te(n, γ)131Te→ 131I U(n, f)→131Te→ 131I	Thyroid studies	6.4.2, 6.4.3
						Renal studies	6.4.4
						Treatment of thyrotoxicosis	—
						Treatment of thyroid cancer	—
127Xe	36 d	EC	—	0.17	133Cs(p, 2p5n)127Xe	Lung function studies	6.4.2
				0.20			
				0.38			
133Xe	5.3 d	e⁻	0.34	0.081	U(n, f)→ 133Xe	Lung function studies	6.4.2
137Cs	30 yr	e⁻	0.51	0.662	U(n,f)→ 137Cs	Calibration source	3.2.2
198Au	65 h	e⁻	0.96	0.41	197Au(n, γ)198Au	Treatment of intrapleural or intraperitoneal neoplasms	—
201Tl	74 h	EC	—	0.07	203Tl(p, 3n)201Pb→ 201Tl	Myocardial imaging	6.4.2

† EC = electron capture; IT = isomeric transition.

Bibliography

Belcher E H and Vetter H (eds) 1971 *Radioisotopes in Medical Diagnosis* (London: Butterworths)

Hine G J (ed) 1974 *Instrumentation in Nuclear Medicine* vols 1, 2 (New York: Academic)

Maisey M 1980 *Nuclear Medicine—A Clinical Introduction* (London: Update)

Subramanian G, Rhodes B A, Cooper J F and Sodd V J 1975 *Radiopharmaceuticals* (New York: Society of Nuclear Medicine)

Thorell J I and Larson S M 1978 *Radioimmunoassay and Related Techniques* (St Louis: C V Mosby)

Early P J, Razzak M A and Sodee D B 1975 *A Textbook of Nuclear Medicine Technology* 2nd edn (St. Louis: C V Mosby)

HPA 1977 *Topic Group Rep.* **16** *The Hospital Preparation of Radiopharmaceuticals* (London: Hospital Physicists' Association)

HPA 1978 *Topic Group Rep.* **27** *The Theory, Specification and Testing of Anger Type Gamma Cameras* (London: Hospital Physicists' Association)

HPA 1980 *Topic Group Rep.* **33** *An Introduction to Automatic Radioactive Sample Counters* (London: Hospital Physicists' Association)

Wagner H N (ed) 1968 *Principles of Nuclear Medicine* (Philadelphia: W B Saunders)

References

Anger H O 1967 *Instrumentation in Nuclear Medicine* vol 1 ed G J Hine (New York: Academic) pp 485–552

Atkins G L 1969 *Multicompartmental Models for Biological Systems* (London: Methuen)

Baker P S 1966 *US Atomic Energy Commission Rep.* CONF-65111 *Radioactive Pharmaceuticals* ed G A Andrews, R M Kniseley and H N Wagner pp 129–42

Bartholomay A F 1968 *Quantitative Biology of Metabolism* ed A Locker (New York: Springer) pp 45–65

Bell T K, Bridges J M and Nelson M G 1965 *J. Clin. Path.* **18** 1

Bender M A and Blau M 1963 *Nucleonics* **21** 52–6

Bentley S A, Glass H I, Lewis S M and Szur L 1974 *Br. J. Haemat.* **26** 179

Bergner P-E E 1961 *J. Theor. Biol.* **1** 120, 359

—— 1962 *Acta Radiol. Suppl.* **210** 1

—— 1964 *J. Theor. Biol.* **6** 137

—— 1965 *Science* **150** 1048–50

—— 1967 *US Atomic Energy Commission Rep. CONF*-661010 *Compartments, Pools and Spaces* pp 21–52

Berman M and Schoenfeld R 1956 *J. Appl. Phys.* **27** 1361–70

Berman M, Shahn E and Weiss M F 1962 *Biophys. J.* **2** 275–87

Birks J B 1964 *The Theory and Practice of Scintillation Counting* (New York: Pergamon)

Bolton A E 1977 *Radioiodination Techniques* (Amersham: Radiochemical Centre)

Boyd R E 1973 *Radiopharmaceuticals and Labelled Compounds* vol 1 (Vienna: IAEA) pp 3–25

Bransom E D 1970 *The Current Status of Liquid Scintillation Counting* (New York: Grune and Stratton)

Britton K E and Brown N J G 1971 *Clinical Renography* (London: Lloyd Luke)

Brownell G L, Ellett W H and Reddy A R 1968 *J. Nucl. Med.* **9** suppl 1

Chantler C, Garnett E S, Parsons V and Veall N 1969 *Clin. Sci.* **37** 169–80

Code of Practice for the Protection of Persons against Ionising Radiations arising from Medical and Dental Use 1972 (London: HMSO)

Cohn S H, Dombrowski C S, Pate H R and Robertson J S 1969 *Phys. Med. Biol.* **16** 645–58

Cook J D, Marsaglia G, Eschbach J W, Funk D D and Finch C A 1970 *J. Clin. Invest.* **49** 197

Cunninghame J G, Hill J I S, Nichols A L and Taylor N K 1978 *Report AERE* R9087 (Harwell: UKAEA)

Dagg J H, Horton P W, Orr J S and Shimmins J 1972 *Br. J. Haemat.* **22** 9–19

Davies D L and Robertson J W K 1973 *Metabolism* **22** 133–7

Deller D J, Worthley B W and Martin H 1965 *Australas. Ann. Med.* **14** 223

Dillman L T 1969 *J. Nucl. Med.* **10** suppl 2

—— 1970 *J. Nucl. Med.* **11** suppl 4

Directory of Whole-Body Counters 1970 (Vienna: IAEA)

Driard B, Roziere G and Rougeot H 1978 *Proc. 2nd Int. Congress* (Washington: World Federation of Nuclear Medicine and Biology)

Eckelman W, Richards P, Hauser W and Atkins H 1971 *J. Nucl. Med.* **12** 22–4

Friedland S S and Zatzick M R 1967 *Instrumentation in Nuclear Medicine* vol 1 ed G J Hine (New York: Academic) pp 73–94

Garnett E S, Parsons V and Veall N 1967 *Lancet* **i** 818

Gray S J and Sterling K 1950 *J. Clin. Invest.* **29** 1604

Grenier J F, Hatano M, Janser J C, Crevoisier R and Weiss A G 1965 *Radioaktive Isotope in Klinik und Forschung* vol 8 (Berlin: Urban and Schwarzenbrug) p 337

Harland W A and Orr J S 1969 *J. Physiol.* **200** 297

Hilditch T E, Horton P W and Alexander W D 1980 *Eur. J. Nucl. Med.* **5** 505

Hine G J 1967 *Instrumentation in Nuclear Medicine* vol 1 (New York: Academic) pp 95–117

Hine G J and Erikson J J 1974 *Instrumentation in Nuclear Medicine* vol 2 ed G J Hine (New York: Academic) pp 1–59

Hobbs J T 1967 *Total Blood Volume—Its Measurement and Significance* (Amersham: Radiochemical Centre)

Horrocks D L 1974 *Applications of Liquid Scintillation Counting* (New York: Academic)

HPA 1978 *Topic Group Rep.* **27** *The Theory, Specification and Testing of Anger Type Gamma Cameras* ed P W Horton (London: Hospital Physicists' Association)

Hutcheon A W, Horton P W, Orr J S and Dagg J H 1977 *Br. J. Haemat.* **37** 195–205

IAEA 1962 *Int. J. Appl. Radiat. Isotopes* **13** 167–72

—— 1972 *Int. J. Appl. Radiat. Isotopes* **23** 305–13

—— 1966 *Technical Report Series No 63 Manual of Radioisotope Production* (Vienna: IAEA)

International Commission on Radiological Protection 1977 *Publication* **26** *Radiation Protection* (Oxford: Pergamon)

Johns H E and Cunningham J R 1974 *The Physics of Radiology* (Springfield: C C Thomas)

Kaihara S and Wagner H N 1968 *J. Lab. Clin. Med.* **71** 400–11

Kenmure A C F, Beatson J McD, Cameron A J V and Horton P W 1971 *Cardiovascular Research* **5** 483–89

Kety S S and Schmidt C F 1948 *Am. J. Physiol.* **143** 53

Kinsman J M, Moore J W and Hamilton W F 1929 *Am. J. Physiol.* **89** 322

Klein O 1930 *Dt. Arch. Klin. Med.* **128** 51

Lassen N A and Kampp M 1965 *Scand. J. Clin. Lab. Invest.* **17** 447

Lassen N A and Munck O 1955 *Acta Physiol. Scand.* **33** 30

Loevinger R and Berman M 1951 *Nucleonics* **9** 26

Maisey M 1980 *Nuclear Medicine—A Clinical Introduction* (London: Update)

Marsaglia G 1963 *Stochastic Analysis of Multicompartment Systems* (Seattle: Boeing Scientific Research Laboratories)

Matthews C M E 1957 *Phys. Med. Biol.* **2** 36

Mills P R, Horton P W and Watkinson G 1979 *Scand. J. Gastroenterology* **14** 913–21

Mollin D L and Waters A H 1968 *The Study of Vitamin B_{12} Absorption Using Labelled Cobalamin* (Amersham: Radiochemical Centre)

Müller T and Steinnes E 1971 *Scand. J. Clin. Lab. Invest.* **28** 213–7

Notes for Guidance on the Administration of Radioactive Substances to Persons for Purposes of Diagnosis, Treatment or Research 1979 (London: DHSS)

Orr J S 1975 *Recent Advances in Nuclear Medicine* ed W A Greig and F C Gillespie (Edinburgh: Churchill–Livingstone) pp 188–213

Orr J S and Gillespie F C 1968 *Science* **162** 138–9

Perl W and Chinard F P 1969 *Science* **166** 260

Radiochemical Manual 1966 (Amersham: Radiochemical Centre)

Rapkin E 1974 *Instrumentation in Nuclear Medicine* vol 2 ed G J Hine (New York: Academic) pp 509–47

Reeve J, Wootton R and Hesp R 1976 *Calcif, Tiss. Res.* **20** 121–35

Richards P 1966a *US Atomic Energy Commission Rep.* CONF-65111 *Radioactive Pharmaceuticals* ed G A Andrews, R M Knisely and H N Wagner pp 155–63

—— 1966b *US Atomic Energy Commission Rep.* CONF-65111 *Radioactive Pharmaceuticals* ed G A Andrews, R M Kniseley and H N Wagner pp 323–34

Richards P, Lebowitz E and Strang L G 1973 *Radiopharmaceuticals and Labelled Compounds* vol 1 (Vienna: IAEA) pp 325–41

Ricketts C, Jacobs A and Cavill I 1975 *Br. J. Haemat.* **31** 65

Riviere R, Comar D, Kellershohn C, Orr J S, Gillespie F C and Lenihan J M A 1969 *Lancet* **i** 389

Robertson J W K, Shimmins J, Horton P W, Lazarus J H and Alexander W D 1971 *Dynamic Studies with Radioisotopes in Medicine* (Vienna: IAEA) pp 199–210

Rowan J O 1981 *Physics and Circulation* (Bristol: Adam Hilger)

Rupp A F 1973 *Radiopharmaceuticals and Labelled Compounds* vol 1 (Vienna: IAEA) pp 223–37

Sandor T, Conroy M F and Hollenburg N K 1970 *Math. Biosci.* **9** 149–59

Schilling R F 1953 *J. Lab. Clin. Med.* **42** 860

Schwabe A D, Cozzetto F J, Bennett L R and Mellinkoff S M 1962 *Gastroenterology* **42** 285–91

Serafini A N and Beaver J E 1978 *Medical Cyclotrons in Nuclear Medicine* (Basel: S Karger)

Sheppard C W and Householder A S 1951 *J. Appl. Phys.* **22** 510–20

Silver S 1962 *Radioactive Isotopes in Medicine and Biology* (London: Henry Kimpton) p 183

Silvester D J 1973 Radiopharmaceuticals and Labelled Compounds vol 1 (Vienna: IAEA) pp 197–222

Smith R F 1980 *Br. J. Clin. Equipment* **5** 100–3

Snyder W S, Ford M R, Warner G C and Fisher H L 1969 *J. Nucl. Med.* suppl 3

Sorenson J A 1974 *Instrumentation in Nuclear Medicine* vol 2 ed G J Hine (New York: Academic) pp 311–48

Strang R, Horton P W and Gillespie F C 1979 *Phys. Med. Biol.* **24** 964–75

Subramanian G, Rhodes B A, Cooper J F and Sodd V J 1975 *Radiopharmaceuticals* (New York: Society of Nuclear Medicine)

Svoboda K 1970 *The Uses of Cyclotrons in Chemistry, Metallurgy and Biology* ed C B Amphlett (London: Butterworths) pp 383–94

Syme D B, Wood E, Blair I M, Kew S, Perry M and Cooper P 1978 *Int. J. Appl. Radiat. Isotopes* **29** 29–38

Thorell J I and Larson S M 1978 *Radioimmunoassay and Related Techniques* (St Louis: C V Mosby)

Turner J C 1971 *Sample Preparation for Liquid Scintillation Counting* (Amersham: Radiochemical Centre)

Vonberg D D, Baker L C, Buckingham P D, Clark J C, Finding K, Sharp J and Silvester
    D J 1970 *The Uses of Cyclotrons in Chemistry*, *Metallurgy and Biology* ed C B Amphlett
    (London: Butterworths) pp 258–69
Whitehouse W J and Putnam J L 1953 *Radioactive Isotopes* (Oxford: Clarendon)
Whittaker E and Robinson G 1944 *The Calculus of Observations* (London: Blackie)
Wicks R and Blau M 1979 *J. Nucl. Med.* **20** 252–4
Widman J C and Powsner E R 1974 *Med. Phys.* **1** 58
Wise M E 1978 *Curr. Top. Radiat. Res. Q.* **12**
Zierler K L 1965 *Circulation Res.* **16** 309
—— 1969 *Science* **163** 491–2

Index